Applied Ethics i

Applied Ethics in Animal Research

Philosophy, Regulation, and Laboratory Applications

**Edited by John P. Gluck,
Tony DiPasquale, and F. Barbara Orlans**

PURDUE UNIVERSITY PRESS

West Lafayette, Indiana

Library of Congress Cataloging-in-Publication Data

Applied ethics in animal research : philosophy, regulation, and laboratory applications / edited by John P. Gluck, Tony DiPasquale, and F. Barbara Orlans.

  p. cm.
  Includes bibliographical references and index.
  ISBN 1-55753-136-6 (alk. paper)—ISBN 1-55753-137-4 (pbk. : alk. paper)

  1. Animal experimentation—Moral and ethical aspects. 2. Laboratory animals—Moral and ethical aspects. 3. Animal welfare—Moral and ethical aspects. I. Gluck, John P., 1943– II. DiPasquale,Tony, 1963– III. Orlans, F. Barbara. IV. Title.

HV4915 .A66 2000
179'.4—dc21   00-062783

Contents

One may ask even the devotee of science that he should acquire an ethical understanding of himself before he devotes himself to scholarship and that he should continue to understand himself ethically while immersed in his labors.

—*Søren Kierkegaard*

Preface

During a recent intense discussion about the implications of considering animals as having moral standing, a prominent biomedical researcher defended her resistance to applying the concept to animals. She explained that she considered animals to be just another category of consumable laboratory "supply," like glassware and computer disks. When asked to expand her position, she pointed out that the supply category was the place in the budget where animal costs were listed and justified in a typical grant application. Although she was clearly attempting to interject a sense of irony into the discussion, her example does illustrate one extreme in the debate, one that sees animals as a form of furniture, there for our use and benefit and deserving of only minimal and indirect ethical concern. At the other extreme are those who see nature to be a virtual democracy, with all or most animals deserving rights that protect against unwanted life intrusions by humans, no matter how beneficial to humans the intrusions might be. Although these extreme positions continue to be forcefully advanced, they have begun to give way to what has been referred to as the "troubled middle." The middle is troubled for a number of reasons. Some question whether the middle is philosophically coherent, whereas others argue that the middle is troubled because it actually confronts the day-to-day reality of human-animal relationships, which seem from some vantages to be indeed irrational and inconsistent. Some have also charged that the debate has been fought primarily with rhetorical devices devoid of validity checks on the substance. Proponents of both sides have shamelessly marched images of dying babies and suffering animals across screens and in newspapers in ways calculated to make one react solely from the gut. The result is that this profound ethical issue is converted to crude counting of which picture provokes more gut reactions.

However, for over twenty years, Barbara Orlans, pioneer in the field of animal ethics, has taken a different tack. Instead of presenting emotional appeals, she has challenged people of good faith to face the issues relevant to the ethics of using animals in biomedical and behavioral research. Toward that end she created a series of conferences specifically focused on the topic of the ethics of animal research. The clear aim of these conferences has been to create a unique venue in which the representatives of the various perspectives in the debate could come together and listen to one another in formal presentations, small group discussions, and casual social encounters. These have been remarkable conferences, attended by scientists, some deeply involved in the use of animals, philosophers representing a number of contrasting theoretical perspectives, animal advocates from across the spectrum of activism, historians of science, and members of the interested public. To date seven of these conferences have taken place, three at the Kennedy Institute of Ethics at Georgetown University (1991, 1994, and 1995), one at the Poynter Center at Indiana University (1996), one at the University of New Mexico Health Sciences Center (1997), one at the University of California at Davis (1999), and most recently, at the Tufts University School of Veterinary Medicine (2001). Feedback from attendees has consistently revealed that the conferences have provided a greater understanding of the issues and an improved respect for the people involved.

What follows in this volume is a collection of essays contributed by individuals who have presented their ideas at these conferences and who fit squarely into the troubled middle. The authors address the issues of philosophy, statutory regulation, and laboratory application of ethics in ways meant to avoid sheer rhetoric and attempts to manipulate. The essays are tempered by open discussion with individuals with different opinions, not merely audiences of true believers. R. G. Frey and Nikola Biller-Andorno expand the notions of the moral standing of animals, Anita Guerrini analyzes the history of the methods of argumentation, Marc Bekoff addresses the implications of what we have begun to know about the minds of animals, Bernard Rollin and Barbara Orlans describe the nature and value of regulatory structures, and David Morton shows how respect for animals can be converted from theory to action in the laboratory. It is the hope of all of us that careful consideration of the positions in these chapters will leave the reader with a deepened understanding and not necessarily with a hardened position. As Bertrand Russell has said, "What is wanted is not the will to believe, but the will to find out, which is the exact opposite."

Acknowledgments

A host of capable administrators aided the conferences that underlie the contributions in this volume. Special thanks go to Mohiba Hanif, Irene McDonald, Dorothy Pearson, Alfredo Aragon, Kenneth Pimple, Sue Heekin, and Susan Grogan. In addition, Tom Beauchamp, Joan Gibson, and Charlene McIver provided strong intellectual support and editorial guidance. We are particularly grateful to Eve Thompson of the Barbour Foundation for their consistent and generous financial support. Finally, we must thank the attendees of these conferences who tested the ideas of the presenters and added to their depth.

Introduction

John P. Gluck and Tony DiPasquale

Human investigations into nature and the incredible array of benefits produced by science have been accompanied throughout history by messages
of caution. These cautions warn us of ethical difficulties and dilemmas
that are revealed in the very acts of discovery responsible for the benefits.
It could even be said that the road to benefits is often paved with harms.
For example, according to an ancient Greek myth, Prometheus's theft of
the secret of fire from Zeus resulted in the benefits that led to creation
of advanced human civilization. Yet in the same story Zeus, in retaliation,
looses Pandora on the human world to produce havoc and pain. The para-
ble of the fall in the Old Testament book of Genesis is another ancient
story that describes the dangers inherent in some forms of knowledge. In
the world of modern literature, Robert Louis Stevenson's *Dr. Jekyll and
Mr. Hyde* and Mary Shelley's *Frankenstein* present similar cautionary notes
about science at times going too far (Shattuck 1996).

Later in history formal ethical documents emanating from places such
as Nuremberg, Helsinki, Belmont House in Maryland, and the U.S. Congress transformed such cautionary messages from subtle philosophical
and literary narratives to guiding principles and formal restrictive regulations. At the very least, these documents (The Nuremberg Code, Helsinki
Agreement, Belmont Report, and the Animal Welfare Act) can be seen as
testimony to the claim that something intrinsic to the conduct of research
can limit the scientist's ethical vision where human or animal participants
are involved. Saint Augustine in the fifth century described this notion as
libido sciendi, the problems produced by the lust to know. The message is
clear: broad yet ethically focused thinking rarely occurs in the context of
full passion. In the eighteenth century the Enlightenment philosopher

1

David Hume counseled skepticism about trusting the validity of the products of reason, arguing that the process is far from independent of self-interest. Hume declared reason to be in fact the slave of the passions. This is not to say, however, that good science and ethically meritorious behavior are contingent on the removal of all emotion.

The Problem of Animals

Animal experimentation poses a special set of ethical problems in the context of research. Although one cannot seriously question that the use of animals in biomedical and behavioral research has contributed to discovery and human benefit, we must also acknowledge that animal subjects have at times suffered serious harms during the process. These harms have emanated from a number of different sources. Central contributors include a certain uncritical acceptance of philosophical positions that exclude animals from ethical consideration in the face of substantive proposals to the contrary, ignorance and misunderstanding about the nature of the lives that animals actually live, and a reluctance to translate what we have come to know about animal lives and perceptions into laboratory procedures and techniques. We will look at these issues briefly in turn, as consideration of these factors forms the foundation of the present volume.

Philosophical Questions

It is generally agreed that the modern resurgence of serious ethical reflection on the moral status of animals began in the 1970s after it had lain relatively quietly since the turn of the twentieth century (see DeGrazia 1991 for an excellent review; Guerrini this volume). The 1970s was a period marked by a general resurgence of the concern for the rights of the oppressed, the expression of antiwar sentiment in the United States, and the women's liberation movement. In essence, these movements questioned the traditional conceptions of where some individuals are ordered in society's priorities and systems of justice. Within this context a small group of pioneering philosophers took up the questions about humankind's relationships with animals and offered critiques of the status quo. For example, Peter Singer's thesis put forth in his important book *Animal Liberation* (1975) turns on the idea that any entity capable of feeling pain, distress, and forms of pleasure must be included under the umbrella of ethical consideration. Singer argued that when an entity is capable of feeling pain in response to our interactions with it, the reality of these consequences must be included in calculating positive justifications for the intervention. In short, animals matter ethically because their pain matters. With this thesis Singer made contact with the utilitarian philosophers of the eighteenth and nineteenth centuries Jeremy Bentham and John Stuart Mill. Singer did not maintain that animals are always off-limits to human use. Rather, he argued that the justification for using animals must reach a standard whereby the use is seen as maximizing benefits. That is, the predicted benefits of the use must outweigh the costs to the animal, and the benefits cannot be achievable in any other way. In analyzing various uses of animals in science, Singer found this standard largely unmet and therefore concluded that the use of animals was unjustified in these cases.

Tom Regan (1983) took a more unyielding abolitionist perspective when he argued that animals have rights not to be harmed by virtue of the fact that they are "subjects of a life." By this he meant that an entity has inherent and not conditional value (i.e., value is not earned). Simply being an animal with a characteristic life course and a set of interests would thus be sufficient to warrant protection. He argued that most animals possess some sense of themselves as entities existing through time. Furthermore, it could be said that they have an interest in their own welfare and rudimentary beliefs about how to carry on a life consistent with that welfare.

Other philosophers, such as Mary Midgley and Steve Sapontzis, added to the developing consensus that animals matter morally. Midgley (1984) developed the thesis that although our tendencies toward social bondedness rightly lead us to prioritize humans—and particularly members of our own social network—ahead of animals in many matters of ethical choice, the status of animals cannot be completely dismissed. Sapontzis (1987) further pointed out that attempts to alter the common practices of society that typically ignore the interests of animals are positively grounded in humans' long-accepted overarching moral goals to become individuals who relieve and reduce suffering and attempt to behave fairly. In this vein Bernard Rollin (1989) challenged the notion that there is a right to research and pointed out the lengths to which some scientists have gone to deny the reality of animal suffering. Although all these positions

certainly contain limitations and inconsistencies, the weight of the conclusions is that the use of animals for science requires serious reconsideration and alteration, with a burden of proof and standards of evidence.

Far from capturing the attention of scientists, however, these positions were in large part ridiculed and ignored (White 1990). In their place
alternative philosophical views that favored the use of animals gained
prominence. The views of the philosopher Carl Cohen in particular took
center stage. In 1986 Cohen wrote a paper entitled "The Case for the Use
of Animals in Biomedical Research," which was published in the influential *New England Journal of Medicine*. The article was then and probably
is still the received position among most biomedical scientists. Cohen
forcefully argued that although we have obligations to animals not to treat
them badly or use them unnecessarily in research, any notion of rights,
no matter how construed, is ludicrous when applied to animals. Rights
for Cohen are valid or potential "trump" cards played against external attempts to alter a person's life plans or preferences without his or her consent. They are trumps in the sense that once they are brought to bear or
claimed by a person, they are absolutely sufficient in negating the ethical
foundation of the intended interference. In other words, a researcher who
has a strong scientific justification to examine the brain of a human patient showing symptoms of an important disorder can proceed with that
examination only up to the point the patient says "stop." This is so even
if the patient had earlier given consent and the predicted benefits of the
examination for the patient and society are exceptionally high. For Cohen, rights are relevant only to members of moral communities, places
in which these types of reciprocal agreements are negotiated and acknowledged. Therefore, his theory of the basis of the right to ethical protection would appear to be grounded in the possession by community
members of a set of cognitive abilities, which gives rise to the creation
and maintenance of true moral communities. These cognitive criteria
seem to require at a minimum such things as consciousness, self-awareness, and the ability to raise questions about one's behavioral motives (so-
called second-order intentions or reflective consciousness). According to
Cohen, these characteristics appear to be uniquely human. Therefore, he
concluded, using animals in research does not violate their rights because
they have none. Instead, we have a moral imperative to do more useful

research with animals, not less. This position seemed to provide a foundation of secure support for an expansive animal-research enterprise.

Since the publication of Cohen's article, others have raised very serious questions about the credibility of his analysis. For example, Beauchamp (1997) pointed out that Cohen neglected to acknowledge the long-accepted relationship between obligations and rights: that is, if the common morality agrees that animals should be exposed only to the minimum amount of discomfort consistent with good science, and that animal alternatives should be used in order to reduce animal numbers and usage, it is clearly appropriate to state that animals have a right to such treatment. Obligations and rights become disconnected when the sense of obligation arises only from a personal feeling of charity and not from a social consensus. In addition, Cohen's analysis would seem to be logically consistent with the use of some kinds of human beings in invasive biomedical research. For example, humans who have permanently lost or never had the ability to experience consciousness or themselves as reflective persons (e.g., anencephallic infants and people in a permanent vegetative state) would seem to lose their membership in the rights-owning moral community. However, Cohen maintains their membership by basically asserting that membership in the human species alone is sufficient to confer rights protection. This would seem then to require that chance evolutionary factors (e.g., reproductive isolation) leading to the formation of species be elevated to the status of having utmost moral significance. This move further requires basically that the Judeo-Christian worldview be accepted as universal. In accepting his arguments without question, members of the research community prematurely raised Cohen's analysis to the level of indisputable, thereby seriously constricting the discussion by scientists affected by the analysis. In the present volume R. G. Frey and Nikola Biller-Andorno refocus and extend this discussion in challenging ways.

Animal Minds

As the previous discussion suggested, many researchers have put a heavy emphasis on the relationship between cognitive characteristics and the level of deserved moral protection. Therefore, one might expect the typical exclusion (or limited inclusion) of animals under the umbrella of moral protection to be based substantially on what we believe animal lives to be or not to be. For example, Peter Carruthers (1992) posed the argument that while many different kinds of animals may seem to behave in ways that look to be versatile, conscious, and creative, their experience may be nonconscious. He asked whether the apparent conscious component in animal behavior is like the person who gets in his car in Albuquerque, New Mexico, and drives the fifty-seven miles to Santa Fe. Upon arrival at the proper highway exit, the driver "comes to" and realizes that he remembers almost nothing about the trip. However, anyone watching the driver from the outside would have noted evidence of supposed consciousness as the person altered speed, changed lanes, made judgments, and avoided danger. If animal behavior is nonconscious in a similar way, then there are no harms to them that we need to ethically consider. While Carruthers cautioned that this is as yet too controversial and insecure a perspective on which to base standards of animal treatment, he accentuated the importance of knowing about the minds and experiences of animals.

In another sense Carruther's notions of nonconscious behavior have found support in the theoretical speculations of many animal researchers throughout a substantial part of the twentieth century. In the tradition of psychological behaviorism, founded by John Watson early in the century, researchers relegated topics such as consciousness, awareness, and imagery to the status of either unessential topics or virtual delusions. Watson (1913) was driven to create what he thought was a truly scientific psychology. From his perspective, elevating behavioral psychology to that level required that it model itself after the successful natural sciences of physics and chemistry. This, he believed, required that the discipline limit itself to those phenomena whose evidence of existence could be reliably agreed to (i.e., directly observable by others). Therefore, scientific discussions of such things as imagery, feelings, consciousness, and emotions would have no place in the science. Animals (and humans) were to be understood from the outside with no reference to their putative subjective lives. Others in the radical tradition went further. Instead of considering these topics as methodologically inaccessible to scientific study, they determined that these states actually did not exist, or at least that they did nothing with respect to causing or influencing behavior. As Rollin expressed the radical behaviorist account, "we don't have thoughts, we only think we do" (1989, 98).

The wide acceptance of behavioral psychology either in its methodological or radical form deferred discussion of the inner life of animals and the moral questions about research with them that would have arisen from such a discussion. Instead, during this period researchers learned a great deal about what animals could do in terms of solving mazes and adjusting to contrived contingencies of reinforcement, but the domain of the experiencing and feeling interior remained empty or at least unexamined.

This situation persisted until the so-called cognitive revolution of the 1960s. In this shift psychologists and ethologists confirmed that the behaviorist program was too costly in its reluctance to study the mental life of humans and animals. Too much was left out. One should remember that the behaviorist move was not motivated by new discoveries that negated the importance of internal lives. Rather, it was a value deemed necessary to pursue a particular vision of science. There hardly could be a doubt that animals have a level of perceptual awareness and are capable of experiencing simple pleasures and pain. In addition, careful observation of animals seemed as well to strongly suggest that many animals are capable of intention: they make basic plans and follow a process of decision making meant to execute the plan. Donald Griffin (1981, 1984, 1992) wrote a series of books that implored the serious researcher to refocus research activities on the question of animal consciousness. He argued that one must not ignore the implications of anecdotal accounts of such things as cooperative hunting in lions, the activities of honey guides, the symbolic dances of bees, and deception in primates. He called for a thorough examination of these possibilities in both field and laboratory contexts. So began a rebirth of interest in animal consciousness.

Where are we now with respect to these issues? We find ourselves in transition, a unique mixture in which the remnants of the behaviorist aversion to the study or rejection of the existence of internal states and the sometimes wildly unguided anthropomorphism coexist side by side. Hauser (2000) has claimed that both extremes are off the mark and emphasized that knowledge of the details of the minds of animals cannot be rejected as nonexistent and cannot safely be generalized from what we know about adult human experience. Unique evolutionary pressures and considerations have shaped each animal species' mind. But while Hauser failed to consider the moral implications of what we are coming to know about the unique minds of animals, Marc Bekoff in his contribution to this volume takes on these implications squarely.

Regulation and Laboratory Applications

Dangers clearly arise when we work with entities whose evolved minds are substantially silent to us, as we may consequently risk underevaluating the suffering brought on by our experimental interactions. However, with active investigation into the cognition and perceptions of animals now established, researchers have begun to recognize that animals have lives that matter to them and that they have interests in pursuing certain characteristic activities as well as in avoiding pain. Now that this has been acknowledged, we must incorporate this information into the design of experiments and development of laboratory procedures.

The history of the development of regulatory standards reflects the recognition of the ethical standing of animals and consideration of the factors that encourage and support their lives as animals. In the United States formal regulation began in 1966 with the passage of the Animal Welfare Act. While this act was at first primarily concerned with regulating the source of laboratory dogs, succeeding modifications of the law have broadened the focus considerably. For example, by the 1970s the law required formal committees to inspect the facilities in which animals are housed prior to experimentation. This meant that the committees spent their energy inspecting the sanitary conditions of the animal holding areas but not the laboratories per se. Formal reports to government and institutional officials scrupulously described such things as the locations of peeling paint, vermin-control procedures, the number of air exchanges per hour in the animal rooms, and the water temperature of cage washers. Thus, the central effort was on parameters only remotely related to the actual welfare of the animals as experiencing animals. Most significant, the committees could not observe the animals during actual experimentation. Members could enter the laboratories only when they were invited to do so by the investigators. This situation led to many gross abuses of animals, primarily out of acts of ignorance and inexperience (Gluck and Orlans 1997). For example, the use of inappropriate anesthesia and analgesics was quite common, as was the use of cage systems that were not linked with the behavioral requirements of the animals they housed. For example, the social primates were typically housed alone in "standard" small cages not even equipped to support climbing and perching activities so clearly important to them. Dogs too were held in single cages with only rare opportunities to exercise and interact with other dogs. Birds such as pigeons were kept in cages that completely precluded the possibility of flying or even fully spreading their wings. In short, housing systems were designed for the convenience of the researcher with little consideration for the behavioral repertoires of the animals.

Due to public criticism and the concern of many scientists, the system of accountability has been vastly expanded (see Orlans 1993, chap. 4). Since 1985 researchers must present their protocols to an Institutional Animal Care and Use Committee (IACUC) comprised of at least five persons including a veterinarian, a scientist experienced in animal research, an institutionally aligned nonscientist, and an individual not associated with the institution in any way. The ethical basis of such a review is grounded in what the English scientists William Russell and Rex Burch (1959) called the "Three Rs." These stand for the goals *replacement*—that is, using animals less capable of experiencing pain or using nonanimal methods whenever feasible—*reduction* of the number of animals to the minimum required to test the hypothesis under consideration, and *refinement* of the experimental design and procedures to limit the distress of the animals to the minimum necessary to accomplish the experimental goals. Therefore, the protocols are expected to outline not only the scientific justification of the experiments but also the ways that the ethical costs to the animals will be reduced or eliminated.

The criticisms of this system have come from both researchers and the public. Public concerns have centered on whether the IACUC can be trusted to provide the objective analysis required or whether it is an example of the fox guarding the hen house. Indeed, a report by the office of the Inspector General of the U.S. Department of Agriculture (USDA 1995) found that IACUCs did not always meet the standards of the Animal Welfare Act. The USDA investigated twenty-six facilities and reported that "twelve of the committees did not always provide assurance that pain and discomfort of animals used in research activities would be minimized" (1995, 24).

Others, such as Stenek (1997), criticized the IACUC system for requiring the membership to accomplish what he considered to be an impossible task. Stenek asserted that in general the members of these committees lack the expertise to evaluate the protocols before them. He considered it to be virtually impossible for members to develop the expertise in animal-care procedures and experimental design necessary to carry out their duties. Therefore, these review activities do little more than divert funds away from needed research activity. In the present volume Rollin develops a report card on the success of the U.S. system of regulation and comes to a different conclusion. David Morton demonstrates in his article that the problem is partially due to the fact that researchers rarely incorporate data-collection procedures that are sensitive to the welfare of the animals as they participate in the experiments. In essence, he provides a training manual for researchers and IACUC members as they review protocols with an eye toward the Three Rs.

The criticisms of the regulatory systems notwithstanding, Barbara Orlans shows that during the last one hundred years the picture from the international perspective demonstrates a steady increase in the presence of national regulations intended to protect animals in experimentation (she puts the number at approximately twenty-three countries). She reports that the parts of the world in which regulation has not yet been developed include South America, Africa, and many Asian countries. Next, she analyzes the nature of the regulations and determines the extent to which the regulations provide protection to animals. She argues persuasively that among other characteristics an ethical system must require the investigator and review committee to specifically estimate the level of the invasiveness of the planned experimental procedures.

Conclusion

The use of animals in biomedical and behavioral research places unique ethical burdens on the scientist. The serious researcher must be familiar with and understand the philosophical considerations of moral standing, the biological and psychological nature of the animals with which he or she works, and the nature of the regulatory guidelines his or her country or jurisdiction has publicly sanctioned. Regardless of the ultimate future of the research ethics issue, the animal researcher will never again enjoy the luxury of a compartmentalized scientific life. Science, ethics, and law are clearly fused, as they should be. Science and ethics are two sides of the same coin, and to behave as if this is not the case invites a visit from Pandora.

References

Beauchamp, T. 1997. Opposing views on animal experimentation: Do animals have rights? *Ethics and Behavior* 7(2):113–21.

Carruthers, P. 1992. *The animal issue: Moral theory in practice.* Cambridge: Cambridge University Press.

Cohen, C. 1986. The case for the use of animals in biomedical research. *New England Journal of Medicine* 315:865–70.

DeGrazia, D. 1991. The moral status of animals and their use in research: A philosophical review. *Kennedy Institute of Ethics Journal* 1(1):48–70.

Gluck, J. P., and F. B. Orlans. 1997. Institutional animal care and use committees: A flawed paradigm or a work in progress? *Ethics and Behavior* 7(4):329–36.

Griffin, D. 1981. *The question of animal awareness.* New York: Rockefeller University Press.

———. 1984. *Animal thinking.* Cambridge, Mass.: Harvard University Press.

———. 1992. *Animal minds.* Chicago: University of Chicago Press.

Hauser, M. 2000. *Wild minds.* New York: Henry Holt.

Midgley, M. 1984. *Animals and why they matter.* Athens: University of Georgia Press.

Orlans, F. B. 1993. *In the name of science: Issues in responsible animal experimentation.* New York: Oxford University Press.

Regan, T. 1983. *The case for animal rights.* Berkeley: University of California Press.

Rollin, B. 1989. *The unheeded cry: Animal consciousness, animal pain, and science.* New York: Oxford University Press.

Russell, W., and R. Burch. 1959. *The principles of humane experimental technique.* London: Methuen.

Sapontzis, S. 1987. *Morals reason and animals.* Philadelphia: Temple University Press.

Shattuck, R. 1996. *Forbidden knowledge: From Prometheus to pornography.* New York: St. Martin's Press.

Singer, P. 1975. *Animal liberation: A new ethics for our treatment of animals.* New York: New York Review.

Stenek, N. 1997. Role of the institutional animal care and use committee in monitoring research. *Ethics and Behavior* 7(2):173–84.

U.S. Department of Agriculture. 1995. *Enforcement of the animal welfare act.* Washington, D.C.: Office of the Inspector General.

Watson, J. B. 1913. Psychology as the behaviorist views it. *Psychological Review* 20:158–72.

White, R. 1990. Letters. *Hastings Center Report* 20:43.

Ethics, Animals, and Scientific Inquiry

R. G. Frey

Abstract: *In this opening chapter, philosopher R. G. Frey sets the ethical questions concerning the justification of animal research in a clear and unrelenting light. He claims that the issue of justification runs deeper than merely the defense for inflicting experimental pain on animals. The foundational ethical issue concerns the use of animals at all, not just the production of distress in animals for the purposes of scientific advancement. Frey requires the reader to evaluate the basis of the claim that we cannot use humans in research in most of the ways that we use animals, even if the humans consented. He asks the reader to consider what characteristics all humans possess that justify their ethical protection but that are lacking in animals and thereby deprive them of similar protection. In attempting to answer this question, he examines the protection-generating ability of such things as cognitive characteristics (i.e., intelligence, sentience, and self-direction), genetic origin, moral community membership, and social and religious traditions. From his perspective, all these criteria fail. However, Frey proposes a relationship between the value of a life and its quality that offers an alternative.*

What, then, are the implications of this analysis for the research enterprise? How are we to proceed from here? Does this analysis require the end of animal research or a lifting of protections from all or some humans? Quite frankly, the implications are extremely controversial.

What may we do to animals in the course of scientific inquiry, whether the primary aim of that inquiry is for our own or their benefit? All too often this question is taken to be about the infliction of pain and suffering upon animals in the course of using them in research. In fact, it raises a deeper issue, not about what justifies the painful use of animals in science, whether for our own or their benefit, but about what justifies their use at all, painful or otherwise. This issue is a deep one, well beyond any simple concern with pain and suffering, however important these may be, and moves us toward undertaking to justify using animals as means to the ends of scientific inquiry (and so, advancement). The question of using animals as means to the ends of scientific inquiry applies to both applied and pure research, to both invasive and noninvasive techniques, and to both painful and painless uses of animals. What needs to be justified is using animals at all, and I assume this need is apparent. For although it is widely held that we may use *animals* as means to the ends of scientific inquiry, it is also widely held that we may not use *humans* to these ends. I do not mean that we may never use humans in scientific research; a good deal of scientific research involves humans. Rather, I mean that we may not use humans in all the myriad ways that we use animals for research purposes, such as models for the production and study of cancers, and we may not do this even if a human per chance consented to be treated in this way.

The deeper issue that my question raises, then, ultimately concerns what justifies using animals in science in ways that would be considered improper to use humans, even humans who consented to the treatment in question.

The Argument against Use

In recent years a general argument has arisen that is widely considered to have made more difficult justifying the use of animals in scientific inquiry, whether it be in applied or pure research and involve invasive or noninvasive techniques. The argument concerns similarities between humans and animals (or, at the very least, the "higher" animals) and in its general form runs as follows. All sides agree that certain features that are characteristic of human beings, such as intelligence, sentiency, and self-direction of their lives, bar using humans in invasive or even noninvasive scientific research without their consent (and in certain types of research even with their consent); these features are present in a great many animals, including the main types of animals used in scientific/medical research; therefore, one is barred from using such animals in much, if not all, such research.

As it stands, this argument might be taken to show the importance of the ability for humans to consent to what is done to them. But the usual significance accorded the argument does not turn on the ability or inability of humans or animals to consent to their treatment. The fact that a rodent cannot consent to what is to be done to it by a researcher, just as numerous humans could not consent to what would be done to them, is not what the argument is about. Rather, it is about the very idea of using creatures who have the characteristics or features noticed, such as intelligence, sentiency, and self-direction, in scientific inquiry at all.

One therefore might construe the argument in its general form to be raising this question: What more must be shown in the case of animals to justify using them in scientific research, given that they share in the characteristics picked out by the argument as the ones around which the case for nonuse of humans is developed?

The Problem of Certain Humans

Some might hold that the whole matter of use could be resolved by simply insisting that animals just do not share in the relevant characteristics singled out by the general argument or do not share in them to the same degree as human beings. Humans, it could be suggested, are just more intelligent, have more numerous and deeper capacities for pain and suffering, and are much more able to direct their lives, and according to their own choices, than animals. But the problem with this proposal—a problem that poses a grave difficulty for those attempting to provide a justification of animal use—is that not all human beings share in these characteristics to the same degree; some clearly share in them to a far less extent than others. What, then, do we do about these humans? If animals lose protection because they fall below the requisite standard of sharing in the characteristics selected as relevant, what about humans who fall below that standard?

A consistent reply here, of course, would be to conclude that, side effects apart, researchers may use those humans who fall below the standard of sharing, just as they use animals, in scientific/medical research. By side effects, I mean such things as the effects on others of researchers using certain human beings in that way. One can easily agree that many people would be outraged by using these humans, who after all could be thought of as the weakest among us, in this way. But side effects have a way of disappearing as scientists undertake educational campaigns to point out the reasons behind what they are doing, and if such side effects disappeared in this case then it would follow that there would be nothing wrong in using some humans in the ways that animals are used.

Even so, the great majority of people would still be strongly opposed to using such humans in scientific research; yet I assume that many of these same people do not find it wrong to use animals in such research. So what can be the difference? What can make it wrong to use humans but right to use animals? An answer to this question is what the argument against use demands from us.

An Assumption about Characteristics

As stated, the general argument for nonuse of animals ultimately depends on a crucial assumption that for any characteristic chosen around which to run the general argument, humans will be found who either lack the characteristic altogether, lack it to a degree that is deemed sufficient to protect them from being used in scientific experiments, or lack it to such a degree that it in fact means some animals have it to a greater degree. For example, few if any will argue with the claim that chimps give evidence of being more intelligent than many severely subnormal humans or that they are sentient to a degree beyond anencephalic infants or that they are better able to direct their lives than humans in the final stages of Alzheimer's disease or senile dementia. Chimps, however, are not unique in these ways; depending on the characteristics selected and thus the humans under consideration, many animals will exceed the human case.

The only characteristic that seems unquestionably to favor humans, regardless of their condition or quality of life, is that of having had human parents. But it is difficult to see why this characteristic is relevant in the important sense. To be sure, in one way it is relevant: it could well be that what we would propose to do to someone in the course of scientific research would upset their parents (and by extension, other family members, friends, and so on). While having parents would obviously matter in this sense, merely being the product of two human parents does not seem in and of itself to matter in a deeper sense: the nature of your parentage says nothing about your present quality of life, your intelligence, your capacity for pain and suffering, your ability to direct your own life, and so on. Indeed, these characteristics seem much more like the things that could serve to distinguish a human life as something that should not be treated as we presently treat animal lives. They address the life being lived rather than what produced that life; they address the nature and quality of the life over which protection from use in scientific research might be extended. Thus, while it is true that anencephalic infants have had human parents, the nature and quality of the lives of anencephalics nevertheless seem by all standards to be far worse than the lives of numerous ordinary animals.

In summary, the general argument against use turns on an assumption that, while able to be overturned, nevertheless appears quite plausible: when selecting any characteristic around which to formulate the argument, one seems inevitably doomed to identify humans who lack that characteristic and animals who to a greater or lesser degree have it.

The Obvious Problem for Cognitive Characteristics

One pressure that this plausible assumption about humans and animals inserts into the argument has been to force those who oppose that assumption to search for characteristics that are highly cognitive in order to ensure the selection of a characteristic that would explicitly bar all animals from the preferred class. Yet any such move seems doomed to failure, since the number of humans who would then fall outside the preferred class would certainly increase, depending on how sophisticated a cognitive task one selects for candidates for potential inclusion in that class. To make the cognitive task less sophisticated in order to encompass as many humans as possible runs the risk that some animals will be included within the preferred class, even as some humans fall outside it.

Thus, a problem emerges in the attempt to find some characteristic (or set of characteristics, including cognitive ones) that separates humans from animals, and that one can plausibly maintain provides a ground for treating humans differently from animals. Do researchers use the humans who fall outside the preferred class in the way they use animals, side effects apart? Or do they protect these humans on some other ground, one that bars the inclusion of any animals in the preferred class and that can plausibly be maintained to anchor a difference in treatment?

The Resort to Abstractions

A powerful temptation now, in order to avoid the characteristics claim, is to opt for something more abstract. For example, one such abstraction is the claim that humans but not animals possess moral rights not to be treated in certain ways. Another is the claim that cultural, social, and moral traditions just do not allow researchers to use humans as they use animals, whatever the characteristics those humans may have or lack, even though those same traditions do allow the use of animals in scientific/medical research. Plainly, any such claim as this quickly runs into difficulty; the cultural, social, and moral traditions under which many people live have permitted slavery or discrimination against women and homosexuals, and the majority no longer view these things as justified. Merely because something is a part of the tradition under which one was brought up or presently lives demonstrates nothing about its moral worthiness as such, even if one may at first glance consider it as morally worthy.

The really interesting point, however, is that a move toward abstraction is not in fact a move away from an appeal to rights to some preferred characteristic (or set of characteristics) at all. The appeal to rights cannot even begin until one specifies the characteristic(s) in virtue of which humans but not animals have moral rights, and much here will turn upon how one construes the cases of those humans who lack the characteristic(s) in question. Here, too, one could try to have recourse to purely "human" rights, but if this does not mean that the operative characteristic is simply held to be having two human parents, then one needs to say what will count as specifically "human" rights. To say that one has a specifically human right not to be tortured seems odd, because it is plain that cats and dogs can suffer immensely and so be tortured. At the very least, if the human right not to be tortured turns in part on the ability to feel pain, then some animals also have this right.

The Resort to Metaphysics

As with the move toward abstraction, the appeal to cultural traditions also does not amount to a move away from the characteristics claim, or so it seems to me.

I regard the central ideas that underlie the claim that our "tradition" bars us from using humans as we use animals to be the following: animals are not members of the moral community, and their lives have no or only little value. Behind these two ideas lies the Judeo-Christian ethic, which holds to a sharp moral difference between humans and animals, and it is clear that these ideas would indeed mark off a sharp moral difference between them. If animals are not members of the moral community, then what one does to them, including invasive research and the infliction of pain and suffering, is not of moral concern; if their lives have no or little value, then the destruction of those lives in the course of research, lives that lie outside the moral community, cannot be of great consequence. The idea of a sharp difference between humans and animals, however, leads one directly back to the characteristics claim and the search for that characteristic or set of characteristics that includes all humans but no animals and that can plausibly be held to anchor a difference in treatment between humans and animals.

Pressure then mounts again on the characteristics claim. As we learn more about animals, especially primates, any sharp break between them and humans becomes more questionable. When 99 percent of a chimp's DNA is the same as a human's, and when they demonstrate an ability to perform feats beyond even some humans, it becomes more difficult to maintain a clear division. Rather, the picture of a continuum of abilities and capacities seems more in order. Again, as we learn more about the massive and progressive deterioration in the quality of some human lives, and find humans in conditions in which they can do little if any of what the chimp can, a sharp break between humans and animals seems even more doubtful. Besides, there is the sheer convenience of it all, of how the Judeo-Christian ethic has advantaged humankind and allowed humans in essence to disregard morally the animate and inanimate environment around them.

Moreover, one can no longer take the religious underpinning of the ideas underlying society's "tradition" as common ground. Today's pluralistic society includes people of many different religions—not all of which agree with the ideas about moral community and the value of animal life—people who reinterpret the Judeo-Christian ethic so that "dominion" over the beasts of the earth means "stewardship" and concern for their well-being, and people of no religious faith. This last group has grown larger, and even whose who proffer a religious ethic tend to do so today in secular terms—for example, speaking of distinctly human goods and human flourishing—that are more amenable to nonreligious people.

The idea that all humans but no animals are made "in the image and likeness of God" no longer seems persuasive as the characteristic that crucially matters concerning treatment, quite apart from any question about the convenience that any such appeal has for using animals for our own ends. The idea that all humans but no animals possess an immortal soul is not even agreed to by all religions, and the thought that all human but no animal life is sacrosanct has similar problems. Yet these attempts to invoke the deity into the argument have a point; for they provide precisely what the characteristics claim so obviously lacks.

What plagues the characteristics claim is that, whatever characteristic (or set of characteristics) one selects for developing the argument against use, one will find humans who lack that characteristic and animals who possess it. The appeal to God had the effect of including all humans within the preferred class of those who could not be used, whatever the other characteristics they lacked: absolutely all humans were made "in the image and likeness of God" or possessed an immortal soul or had lives that were sacrosanct. Precisely what the appeal to God accomplished, then, was to provide a ground for the nonuse of humans in science and medicine by showing that all humans, whatever their condition and their quality of life, did indeed share in a characteristic that animals did not: they were, if I may put it this way, God's "preferred creature." Precisely what is in doubt, in our more secular age, however, is this preferred status.

Secular Alternatives

The problem is that, religion apart, nothing seems to ensure that all human life, whatever its condition and quality, but no animal life, whatever its condition and quality, falls into the preferred class of nonuse. Secular attempts to replace God as the guarantor of the preferred status of humans have not proved successful, and they inevitably involve the search yet again for some magical characteristic that separates humans from animals and that can be a plausible candidate on which to hang a difference in treatment.

For example, the appeal of one secular alternative is doubtless widespread: Why can we not just show partiality for our own kind? The problem here is knowing exactly how to understand "our own kind." Presumably, were I to show in various situations partiality toward individuals of my own race or gender, an outcry would ensue. How then, are we to decide which human characteristics—race, sex, disability, intelligence—are the ones toward which we may show partiality?

But why not species partiality? Why can I not just prefer members of my own species over members of other species? Plainly, I can. But is having this preference morally required of me? Suppose one can save either one's faithful cat who has rendered long and valuable service or some human whom one does not know. If one saves the cat, has one done something wrong? Is there a sense in which one must prefer human beings over animals if one is to be moral?

Suppose now that the human being suffers from a series of terrible maladies that ensure a very low quality of life, together with a prognosis of a much-reduced life span. Must one still prefer the human to the animal? What if the human were an anencephalic infant? Again, these questions seem to be asking for the characteristic or set of characteristics by which we will save life, whether human or animal, in circumstances in which the animal and not the human may best exemplify that characteristic or set.

Another secular alternative that attempts to place all humans in the preferred class is found in Tom Regan's (1982) claim that whatever the quality of human lives, all human life has equal inherent worth. Unfortunately, Regan does not make clear what inherent worth is, how one can recognize its presence in something, and what criteria of identity one should use to distinguish different occurrences of it in quite different sorts of things (for example, humans, dogs, and ecosystems). The really interesting feature of Regan's claim, however, lies in the fact that it is ultimately an attempt to supply a secular analogue to what in the Judeo-Christian ethic was supplied by the claim that all human life is equal in the eyes of God. Clearly, not all human lives have the same quality (whatever may be true of worth); no one in the final stages of amyotrophic lateral sclerosis or in the final stages of pancreatic cancer would say otherwise. Yet in the Judeo-Christian ethic such lives are of equal value to ordinary human lives because they are held to be equal in the eyes of God.

If, however, God is taken out of this argument, then what underpins the claim of the equal value (not quality) of all human lives? Indeed, because the quality of human lives can vary enormously, those who see the value of human life as bound up with its quality will plainly not regard all human lives as equally valuable. To answer this, Regan invents the notion of equal inherent worth, a concept that differs from that of the value of a life such that lives of unequal quality and value can nevertheless have equal inherent worth. Regan provides no argument for this notion, even as he provides no argument for how one can recognize it or even understand what it is.

If one is religious, there seems no need to opt for Regan's secular alternative. If one is not religious, then what underpins a claim of the equal inherent worth of all human lives, including lives of such drastically low quality that even those living those lives often seek some release from them? So far as I can determine, nothing performs this task.

The Preferred Class of Nonuse

The issue of whether one may permissibly use animals in scientific research therefore faces a serious problem from the outset if one simultaneously holds that one may not permissibly use humans in such research; for then one must ask what it is that creates the difference between the human and animal cases. The idea that one can make sharp distinctions via a characteristic or set of characteristics that encompasses all humans but no animals encounters problems. The retreat to more cognitive tasks and abilities will not work. The retreat to abstractions in order to place all humans into the preferred class of nonuse also has difficulties.

A central concern thus emerges: lives of massively different quality may be thought equal in the eyes of God, but with God out of the picture nothing supplies a similar sense of equality. All that is left are lives of different and often radically different quality, and on a quality-of-life view of the value of life, this means that some human lives are more valuable than others. This in turn leads to what for many is a troubling possibility: some human lives can be of a quality and value so low as to be exceeded in quality and value by (some) animal lives, in circumstances in which nothing appears to guarantee that human lives of any quality, however low, are barred from scientific use, while animal lives of any quality, however high, are free to be used. Nothing appears to guarantee that all human lives are in the preferred class of nonuse because nothing guarantees that all human lives, whatever their quality, exceed in value all animal lives, whatever their quality. It is with this result, I think, that any justification of the use of animals in scientific/medical research must begin, and this is where my own attempt at justification has begun, as I have tried to show elsewhere (Frey 1996, 1997, 1998).

Most people will undoubtedly find this starting point unpalatable, but I know of nothing that enables us to avoid it. A justification simply must be given for why it is that humans but not animals attain the preferred class of nonuse, whatever their condition or quality of life.

Rejecting the Underlying Ideas of the Claim of Tradition

Finally, it should be clear that this starting point for discussing the permissible limits of use centrally involves the rejection of the two claims that Western religious traditions have bequeathed us: namely, that animals are not members of the moral community and that their lives have little or no value. Both these theses, I believe, are false. I have space for only a few remarks on each thesis here.

Membership in the moral community is, I think, a matter of whether a creature is an experiential subject, with an unfolding series of experiences that, depending on their quality, can make that creature's life go well or badly. A creature of this sort has a welfare that can be enhanced or diminished by what we do to it. Hence, such a creature has a quality of life. Pigs, rodents, rabbits, and chimps are all such creatures. They are experiential subjects with a welfare and a quality of life that our actions can enhance and diminish, regardless of whether they are creatures with moral rights or capable of moral agency. Accordingly, a creature can be a member of the moral community even if it lacks moral rights and is incapable of moral agency. All it requires is that it be an experiential subject with a welfare and a quality of life, and while we may be uncertain whether some creatures have experiential lives, pigs, rodents, rabbits, and chimps are not doubtful cases. These creatures are members of the moral community in exactly the same way that humans are. All have unfolding experiences and thus are creatures with a welfare and quality of life, and what is done to them affects that welfare and quality of life.

On this view, although pain and suffering may have an important effect on a life, they are simply part of experiential lives on par with the creature's other subjective experiences. Inquiry into the nature of a creature's subjective experience therefore becomes an important part of the inquiry into who or what is a member of the moral community and thus how it may be treated.

With a welfare and a quality of life, it follows that animal life has value, where the value of a life is a function of its quality. All experiential creatures, not just humans, have a welfare and a quality of life that our actions can affect positively or negatively. Quality of life therefore determines the value not only of human but also of animal lives, and quality of life, I think, is a function of the scope and capacities of a creature for different kinds of experiences. It may be true that normal adult humans outstrip animals in these regards, but it is also true that some perfectly healthy animals outstrip some humans in these regards. Hence, the quality-of-life view of a life's value denies that all human lives have the same value, that all human lives have more value than all animal lives, and that there is something that ensures that no animal life, however high its quality, will be more valuable than any human life, however low its quality. Plainly, the quality-of-life view in the present argument is not the functional equivalent of God in the earlier argument for nonuse.

The attempt to justify the use of animals in scientific inquiry must therefore begin by accepting what Western religious traditions have on the whole denied, namely, that animals are members of the moral community, that they have lives of value, and that they on occasion can have lives of higher value than some human lives.

References

Frey, R. 1996. Medicine, animal experimentation, and the moral problem of unfortunate humans. *Social Philosophy and Policy* 13:181–211.

———. 1997. Moral community and animal research in medicine. *Ethics and Behavior* 7:123–36.

———. 1998. Organs for transplant: Animals, moral standing, and one view of the ethics of xenotransplantation. In *Animals and biotechnology*, edited by A. Holland and A. Johnson. London: Chapman and Hall.

Regan, T. 1982. The nature and possibility of an environmental ethic. In *All that dwell therein*, edited by T. Regan. Berkeley: University of California Press.

Can They Reason? Can They Talk? Can We Do without Moral Price Tags in Animal Ethics?

Nikola Biller-Andorno

Abstract: *In this chapter, Nikola Biller-Andorno, a physician and care ethicist, challenges a foundational premise in ethics, and animal ethics in particular. Biller-Andorno questions the defensibility of the notion that if a being is not capable of suffering, or of experiencing enjoyment, then there is nothing to take into account ethically with respect to its life. In other words, is the ability to suffer a sufficient requirement for moral considerability? She then critiques Frey's conceptualization that the degree of ethical protection is tied to the value of an entity's life and that value is based on a life's quality. She asks difficult and important questions when she wonders how differing qualities can be measured and compared. What vantage point should the assignor of value take—his or her view or the view of the animal?*

In challenging these notions, Biller-Andorno creatively applies the idea of the Umwelt, or the environment as it is perceived by an animal, which was originally developed by the ethologist Jakob von Uexkuell in the early part of the twentieth century. Biller-Andorno asks what moral significance there is if we humans are part of the perceived worlds of some animals. Using an expanded concept of empathy, she then proposes that the focus in animal ethics should move away from the question "Who is worthy of protection?" to "Who is in need of protection?" For her, the issue focuses more on instrumentalizing and intruding on the lives of other animals, not merely on whether pain is produced as a consequence.

In the last decades, research in psychology and ethology has confronted us with mourning elephants, altruistic dolphins, and chimps joking with humans in sign language (Blum 1994; de Waal 1996)—compelling images, even if uncertainties remain concerning the interpretation of animal behavior and cognition. At the same time, the emerging discipline of bioethics and a number of well-publicized cases have helped to foster our awareness of medical conditions like PVS (persistent vegetative state) or anencephaly, in which patients lack what are usually regarded as key human characteristics, such as self-awareness and the ability to reason, speak, and relate to fellow humans.

Given this situation, the traditionally stark contrast between the moral treatment of human beings and that of animals became increasingly difficult to justify, especially as religious and metaphysical systems had lost much of their moral authority in pluralistic societies. The special moral status of human beings and the sanctity of human life was no longer beyond doubt, and the subordinate position of animals in the realm of the moral no longer appeared necessarily self-evident. There seemed to be a need for a criterion that would justify setting humans and nonhumans morally apart. Upon closer examination, this proved not an easy task. This difficulty can be formulated as the so-called argument from marginal cases: if the criterion for moral standing is pitched low enough to include all human beings, it will also include a large and diverse group of nonhuman animals; if it is pitched high enough to exclude all nonhuman animals, it will also exclude some human beings (see Callicott 1995, 678; Regan 1983; Singer 1990).

To settle this dilemma, Peter Singer suggested following Jeremy Bentham[1] regarding the ability to suffer rather than being rational or being human as the criterion for moral status: "If a being suffers, there can be no moral justification for disregarding that suffering, or for refusing to count it equally with the like suffering of any other being. But the converse of this is also true. If a being is not capable of suffering, or of enjoyment, there is nothing to take into account" (Singer 1990, 171). Neglecting "the interests of members of other species with equal or superior capacities" (Singer 1995a, 151) would be speciesism, a moral failure Singer compared to racism or sexism. The ensuing debate, joined by proponents of animal rights or welfare, as well as biomedical scientists, philosophers, and ecologists, was lively and often ferocious, sometimes even violent. Opinions differed widely as to where and how the crumbling walls of the realm of moral considerability should be re-erected. Whereas some authors tried to conserve the traditional status quo, many others aimed to widen the scope of ethics to include animals, or even plants, species, and ecosystems, by virtue of their utility for humans, their intrinsic value and corresponding rights, or their contribution to the greatest happiness for all.[2]

Although many issues remained controversial, a good part of the discussion seemed to be carried by a certain confidence that an objective, impartial, and universal perspective could be reached once speciesism had been branded an invalid moral argument and that a fair determination of the moral value of the being or species under consideration was indeed possible.[3] I shall argue that this is not the case. Although I do not doubt that extending our moral scope to include nonhuman beings is necessary for a timely ethical reflection and practice, it is important to realize that we will always remain bound to some degree to our human point of view—as will our attribution of value. This does not mean we cannot make moral decisions. But there is a tremendous difference between striving for neutrality and claiming it: criteria, such as rationality or the ability to feel pain, that might be useful in locating the need for protection and moral concern can all too easily be turned into moral price tags, fixed on other beings by self-proclaimed human judges.

The use of animals in research appears particularly suited to illustrate my concerns about claims to species neutrality and the ascription of moral value and to sketch an alternative approach. This is a problem of considerable theoretical as well as practical importance and priority, cutting across disciplines as well as the usual boundaries between the professional, public, and private. In addition, the emotional and highly controversial quality of this topic can be interpreted to reflect our struggle with an increased moral sensitivity and a fairly unparalleled freedom to act upon it. At this point, many of us are just not sure how to weigh the benefit for frail humans and the cost for healthy animals. On the one hand we are pulled toward the goal of alleviating human—and possibly our very own—suffering, while on the other we wonder if and how such an unequal cost-benefit distribution can be justified, given that many animals used in research carry the criteria that we have come to associate with moral significance. One possible but alarming solution to this dilemma is to justify an experiment by conducting it "on an orphaned human being at a mental level similar to that of the proposed animal subject" (Singer 1995b, 36) to make sure that interests are being weighed in a nonspeciesist way. This suggestion represents the very move that I consider highly problematic: suffering is no longer used as an indicator of an individual's need but as a capacity that confers value to the class of beings that possesses it. This move is not an uncommon one in bioethical argumentation.[4] Some, such as Raymond Frey's quality-of-life view on the value of life, even use a positive criterion to confer and compare such value (discussed in the previous section). I will first use his concept as a referent for my position, analyzing what I perceive as its problems and dangers. I will then propose an alternative empathy-oriented approach for the integration of animals into our moral considerations. Finally, I will demonstrate the implications of such an empathy-oriented approach on a theoretical as well as a practical level.[5]

My focus on the argumentative structures used in the debate on animal experimentation and on our possibilities and limitations as human moral agents implies that my interest is mainly a methodological, or meta-ethical, one.[6] I therefore do not aim to present the contents of this debate, nor do I investigate here the necessity of research on animals or present a definitive position. Rather, my goal is to identify a line of argument I have found to be prevalent in the discussion on the moral status of animals and which I consider both faulty in theory and dangerous if put into practice.

Does Quality of Life Equal Moral Value?

Although the focus on the question "Can they suffer?" has been instrumental in initiating and sustaining the recent debate on animal ethics, the principle of equal consideration of suffering, advocated by Singer (1990), does not suffice for a comprehensive account of our moral treatment of animals. Significantly, it does not give any reason for prematurely ending another being's life if it happens in a painless way and does not contradict any plans for the future.[7] However, if one accepts the notion that human beings can be mistreated without conscious suffering involved, which probably everyone except a strict act-utilitarian would concede (imagine someone being raped under anesthesia, not feeling pain or being aware of or remembering the event), then nonhuman beings without awareness and sentience can be wronged as well, at least from a nonspeciesist point of view.

This is exactly the concern Frey raises in his contribution, "Ethics, Animals, and Scientific Inquiry," in this volume: How can the use of non-human animals in research, no matter if painful or not, be justified if humans are categorically exempted at least from some of these experiments, although they can be expected to harvest the main benefit from the experiments? Dismissing various attempts that resort to tradition or to metaphysical or teleological assumptions, Frey concludes that no such justification is possible given the lack of a characteristic that would set all human beings morally apart from all nonhuman ones. Instead, Frey suggests a "quality-of-life view of the value of life," turning the negative principle of equal consideration of suffering into a positive principle of equal consideration of quality of life. Hence, the lower a being's quality of life—which is taken to be an objectively determinable and commensurable unit—the lower its value and the easier to justify its use for research purposes.

I agree with Frey's analysis that an exclusive focus on suffering does not exhaust the ways in which animals need to be considered morally. Why should only their pain matter, and not the mere fact that they are being used for someone else's advantage? I also concur that trying to explore and understand other beings' subjective experiences is a valuable methodological tool to localize moral conflicts, although I do not think it tells us anything about the moral value of these beings. Finally, I have no doubt that animals have to be considered morally and that preferring humans, in general as well as in specific situations, cannot go without further explanation and justification. Nonetheless, I do not support the implications of Frey's conclusion that in order to avoid speciesist bias we should regard a being's quality of life as an indicator of its value, which means that the most deprived could be most justifiably used for research. If I understand Frey's argument correctly, however, it is a utilitarian answer to a Kantian question:[8] namely, it is easiest to justify using those living beings as means to some end we deem worthy that least enjoy their life (and therefore have not much to contribute to the overall balance of pleasure over pain in the world, I assume).

It is not unusual for utilitarian accounts to arrive at solutions that contrast sharply with prevailing notions of morality.[9] This fact alone, of course, cannot count as an argument against such an approach. It does provide an incentive, however, for closer scrutiny. Frey's notion that it would be only consistent to say that "researchers may use those humans who fall below the standard of sharing [in the characteristics selected as relevant] just as they use animals in scientific/medical research" (Frey, current volume) is not only "unpalatable," as he observes, but runs counter to the central ideas of many religions as well as to our secular human rights. The fact that "the quality-of-life view . . . denies that all human lives have the same value" (Frey, current volume) unavoidably leads to discrimination against those lives. On a professional level, using human beings in research from which they are not likely to benefit, and which entails great risks, clearly conflicts with current guidelines on human experimentation and professional medical conduct. Finally, the use of disadvantaged humans goes against the grain of our common understanding of justice: Are those people—for example, patients with severe Alzheimer's—supposed to suffer even more, instead of being compensated for their loss by special care?

Frey seems to be well aware of these consequences: "[W]e must face this prospect [that the lives of some perfectly healthy animals have a higher quality and greater value than the lives of some humans] with all the implications it may have for the use of these unfortunate humans by others, at least if we continue to justify the use of animals in medical/ scientific research by appeal to the lower quality and value of their lives" (Frey 1988, 196). Yet from his position as a provivisectionist and utilitarian who considers the charge of speciesism a serious one, he sees his "greater value thesis" (Frey 1996, 202) as the only alternative to the abolitionist stance that has resulted from the principle of the equal consideration of suffering. Balancing these options, Frey would rather sacrifice some humans with a low potential for enrichment than forgo the potential benefits of experimental medicine, which he seems to hold in high regard indeed.[10]

I do not believe that the current attempt to integrate animals into our scope of moral concern provides such an either-or situation. I also do not consider the rather repelling consequences that Frey proposes as unavoidable. I propose instead a critical look at the assumptions, argumentations, and claims of the "quality-of-life view on the value of life." First, it is necessary to clarify how "quality of life" is to be understood here. Who, for instance, are those "defective humans" whom Frey suggests using for experiments on retinas rather than "rabbits or chimps" (1988, 196)? One could think of seriously brain-damaged patients, of anencephalic babies, the comatose, and even the severely demented or depressed. It seems to be these kinds of humans to whom Frey would ascribe less value (see Frey 1988, 195);[11] their lives are "less rich of content" (1996, 203), and they have lost their "agency" (1988, 200), the prospect of integrating . . . [their] lives into wholes" (1996, 208), which usually sets humans apart from nonhumans. But what about a human being with a subjectively high quality of life who has no potential for further enrichment due to an impending devastating neurodegenerative disorder? Or what about a healthy newborn baby with a rather limited actual richness of content, but with a considerable potential to acquire it in the future? It is not clear if Frey's notion of quality of life refers to a life's current state of richness of content, to its potential for (further) enrichment, or to both.

Moreover, I think a normative concept of quality of life yields even more substantial problems than the counterintuitive character of its consequences and a certain vagueness in its definition. Although space does not permit a comprehensive account of the potential and limitations of the quality-of-life concept, I will raise here some questions and concerns on a (1) methodological, (2) content, (3) conceptual, and (4) pragmatic level.

1. The following are among the basic methodological questions that arise from a quality-of-life approach: How can the quality of life of different individuals, or even species, be transformed into comparable units so that it can be used to weigh the values of beings against each other? What is the reliability and particularly the validity of such a concept? In other words, can the results be replicated and does the instrument measure what it is supposed to? These questions are of special importance because the measure of a life's quality is supposed to function as an indicator of its value, which in Frey's account is a real "life-or-death" issue. I do not think the instrument is advanced enough—and I doubt it will ever be—to be suitable for such a strong argument in normative deliberations.

2. Questions regarding the content of the quality-of-life concept center mainly around the potential discrepancies of a subjective versus a supposedly objective account. Does this variable of quality of life tell us something about how good and valuable the individual perceives his or her life or about how others judge it? Is it a subjective or an objective measure or a combination of both (see Frey 1996, 205)? In the

case of a combination, what if the two conflict—does the subjective trump the objective or vice versa? Can someone's life be of a lower quality than the person in question takes it to be? And what is a rich inner life, if not a subjectively rich inner life? What is a valuable life, if not a subjectively valuable life? The limitations of our current knowledge and our human epistemological abilities regarding the inner experiences of other, especially nonhuman beings, seem to preclude any claims to objectivity.

3. Some major conceptual concerns arise from the attempt to employ the notion of quality of life in a utilitarian framework. One is the claim to impartiality in the determination and weighing of the quality of life of different beings. How can we avoid having our human perspective influence how important we consider one contribution to the quality of life compared with another, for example, "agency" versus the ability to fly (without technical assistance)? Frey himself questions whether his argumentation may be indirect speciesism "in that, in order to determine the quality and value of a life, I use human-centered criteria as if they were appropriate for assessing the quality and value of all life" (1988, 199). Unfortunately, he does not, as far as I can determine, address that charge beyond asserting that his concept is preferable to the alternative of considering all life equally valuable. I think his attempt "to gauge the quality of their (animal) lives in terms appropriate to their species" (Frey 1996, 205) can yield only species-specific qualities of life: the quality of life of a mouse has to be judged by mouse standards, and I do not see how those can be compared with goose standards; so how can a common denominator result? This leads to the important charge of incommensurability of the quality of life of different species, if not of different individuals within a species, which Frey denies, again without offering an argument that can explain why experiences should in fact be comparable (207).

4. Finally, problems exist on a pragmatic level, that is, concerning the relevance of the concept for moral acting and decision making. First, the connection between quality of life and value of life is not as straightforward as it might seem at first sight. From a subjective understanding of the notion of quality of life, one can argue that someone can think his or her life is going miserably, poorly, but can still value it. And if quality of life is considered to have an objective

character, objections also arise against a *universal claim* to deduct the value of a life from its quality independently from the moral agent's spectrum of possible actions and the necessity to make a concrete moral decision. Furthermore, from a deontological view, one can protest that by ascribing value to a being it is already being seized, instrumentalized; it is no longer regarded as an invaluable end in itself but is given a price—something that Frey had set out to avoid. Finally, from a utilitarian position, one could object that the quality-of-life concept seems to look at only one side of the coin: the more potential for enrichment, the more value. But isn't it also the case that an already rich life or a potential for enrichment also comes with a corresponding potential for "misery," in the sense of a decrease of these high standards?

Given all those reservations about a descriptive, but particularly a prescriptive, concept of quality of life, I am not convinced that an objectively, impartially, and universally determined value of life is possible. I would rather call such descriptions "pseudo-qualities," qualities one would like to be true because they seemingly make moral deliberations so much easier, but which in fact dangerously cover the distinction of facts and normative definitions. Thus, I do not believe that the quality-of-life approach represents a nonpositioned, neutral perspective, not with regard to species and probably not even with regard to groups or individuals within a single species.

I do not intend to introduce a relativist stance, nor do I want to recapitulate noncognitivist reservations toward the possibility or truth of normative statements. But I do want to point to the danger of reintroducing or reinforcing discrimination under the cover of unjustified implicit claims. According to Frey's "quality-of-life view on the value of life"—in which there can be no one-fit-for-all quality of life for the whole species—wouldn't a well-off, healthy philosopher have to be regarded as more valuable than a poor, distressed factory worker, whose life is less rich in content and who is less able to mold his life as he chooses? In order to avoid this rather appalling consequence, the quality-of-life concept in my view has to be limited to the subjective perspective of the human being or nonhuman being in question; this would, however, render its results incommensurable and thus useless for utilitarian calculations.

But even supposing the quality-of-life concept could ever be turned
into a precisely determinable and commensurable measure, would it
really have much relevance as a moral term? One of the most fundamental concerns of ethics is the avoidance of harm, but having a low quality of life does not necessarily mean being less vulnerable. Shouldn't moral
considerations focus more on how to protect those vulnerabilities than
on how to prevent rich lives from getting poorer? I see a legitimate role
of the quality-of-life concept only in the exploration of a being's subjective situation and its preferences in order to enable the moral agent to
avoid harm and foster well-being. I believe we need another approach to
guide us in moral decision making, one that focuses on vulnerabilities
rather than positive values and refrains from prima facie ascriptions of
value. In the following sections, I sketch some features of such an attempt.

Rich Life or Not: Who Cares?

Singer's principle of equal consideration of suffering aims to extend the
realm of ethics to include nonhuman vulnerabilities and to fight discrimination based on species membership (see Singer 1990, 243). Shifting
the questioning this way from "Do they meet a certain criterion?"[12] to
"Can they be harmed?" is an important move but does not completely
avoid the speciesist trap. The notion of suffering is open to an abundance
of interpretations, which are all modeled to some degree on human capacities and experiences and whose recognition in other beings is dependent on our limited human perception. Environmental ethicists in particular have criticized this "extensionist approach" (Hargrove 1992, xxiv),
which confers value on the basis of characteristics shared with humans.[13]

It thus seems reasonable to expand the question of the moral per-
missibility of animal experimentation to ask whether we may use animals
regardless if it involves any form of suffering. One such way to proceed
is to make the value of all life dependent on one criterion, such as quality of life in Frey's case, and to justify instrumentalizations by creating a
hierarchy of individuals according to this criterion. This means, however,
that the determination of moral value once more gains precedence
over the search for beings in need of protection, thus reopening the
door to discrimination. "The more (quality of life) you have, the more

(protection) you get" is the obvious, although perhaps unintended, message. If on the other hand someone's quality of life was low (for example, having lost part of his cognitive abilities, affect, senses, and memory, not an uncommon occurrence after a bad stroke), this person's "value" would be "down," and in addition to all that misery, he would find himself short-listed to be used in research for the good of those who are better off. This is of course a drastic, but I believe not incorrect, formulation of the implications of such an "unequal value thesis" (Frey 1988, 192), which frankly undermine the most basic of our humanistic ideals.[14]

Another approach is to ascribe inherent worth to nonhuman beings, which renders their use as a means an intrusion that has to be justified.[15] In order to explicate the source of this attributed value, however, some kind of religious or metaphysical assumption has to be invoked, which makes these approaches attractive only to those who are willing to share these assumptions. I will attempt still another nonutilitarian approach, which refrains from any a priori ascription of value and focuses on the need for protection or help rather than an—I believe necessarily presumptuous—assessment of another being's value and worthiness.

I want to shift the question from the passive "Can they be harmed?" further to "Can we harm them?" thus reconnecting moral reflections on the "moral patient" with the situation of the moral agent.[16] In general, the identity of the moral agent today is shaped by three features. First, Kantians and non-Kantians alike would probably agree that the ability to impose limits on ourselves, to act autonomously, is one of the central features of human moral agency.[17] However, the fact that nonhuman animals are not autonomous beings does not imply that they exist only as means for human purposes (U. Wolf 1990). The imperative to use a being "as an end, never merely as a means" (Kant [1785] 1994, 429) does not need to be limited to persons unless reason is considered the only absolute value. Second, through advances in science and technology, many members of our society have unprecedented freedom in thought and action and power to choose from an array of options—to the extent that science and technology themselves have become objects of critical inquiry. Third, due to this ubiquitous influence of human action, we are in some form part of the environment of probably every living being on this planet. Conversely, many other beings are, directly or indirectly, part of our subjective environment.

Who's In? On the Scope of Moral Considerability

It seems plausible to assume that an entity that exists within a subjective environment and sustains its existence may do so without any further reason and to place the burden of justification on any intruder that has the ability to reflect on its action and has a choice of options. Indeed, there are no good reasons why any living being should have to justify its existence by serving another being's needs (see U. Wolf 1990, 75). If we claim freedom from such justification for our own species, we have to concede it to others, too. Prima facie including all living beings into the realm of "moral considerability" (Goodpaster 1978, 316) also helps reduce arbitrariness as well as speciesist bias, which we can hardly avoid when we pick a more exclusive criterion for membership to the realm of moral relevance and try to apply it to beings of other species.

Nonhuman beings are not individual black boxes but have subjective environments or *Umwelts* that we can enter and explore. This fact is the epistemological basis for the possibility to extend ethics beyond personhood. The term *Umwelt* is borrowed from the biologist Jakob von Uexkuell (1864–1944), who, strongly influenced by Kantian philosophy, tried to avoid the pitfalls of both mechanism and vitalism. His work has inspired psychosomatic medicine (see T. von Uexkuell 1997), as well as generations of ethologists, among them Konrad Lorenz and Niko Tinbergen. His definition of life centers on the active creation of an individual environment out of neutral surroundings via a living being's perceptual and operational organs, which appropriate information from outside and assign meaning to it (see J. von Uexkuell 1926). Every subject and its environment form an integrated, dynamic entity. The *Umwelt* is constantly being modified by external influences as well as the individual's own action, which in turn changes its sensory perception.[18]

Therefore, we as human moral agents are potentially or actually part of all other beings' subjective environments in a constructive, neutral, or destructive way. Furthermore, living beings may have mechanistic properties, but beyond that they are also in possession of a self-directed creativity. We can thus wrong other beings by unjust instrumentalization, regardless of whether it directly hurts their physical integrity. Our moral relationship to nonhuman beings therefore has a justice perspective as well as a care perspective. This does not force us, of course, to conclude that we may never use any other being as means or that we have to care for every being all the time. But it requires us to generally acknowledge the moral relevance of our role as parts of other beings' environments and to balance our interests with the moral imperative to minimize instrumentalization and harm.

I have deliberately phrased the assumption that all living beings are ends in themselves—in the sense that they do not have to justify their existence (to us humans) by serving as a means to some other purpose—in a way to avoid any strong claims such as the existence of intrinsic value (Fox 1993), teleological centers (Taylor 1986), or a will to live (Schweitzer 1989). I realize the danger of projecting interests or preferences onto beings where there may be none. On the other hand I think the insistence that interests are bound to some minimal requirements, such as consciousness or sentience (see Feinberg 1974; Dworkin 1994), can unduly limit our moral scope from the very start. Anthropomorphist restrictions can be just as inadequate (e.g., a human notion of suffering) as anthropomorphist extensions of moral concerns. I believe we should err on the side of the latter—*in dubio pro reo* (in doubt for the accused). I would claim respect for another being's end, "equally whether it can express itself to my comprehension or whether it remains unvoiced," as Schweitzer puts it (1989, 32–33).[19]

Yet our abilities for moral treatment of other beings are bound to our knowledge about them. As J. C. Wolf explains, "We need empirical data on the nature of those affected by our decisions and have to be able to project ourselves into their situation—otherwise, the realization of the consequences of our action will hardly motivate us to moral consideration" (J. C. Wolf 1992, 174, translation by the author). Exploring other beings' subjective environments and analyzing our factual and potential roles in those environments can help us extend our knowledge base for moral reflections and decisions with regard to nonhuman beings. The individual subjective worlds are our only access to reality, not some general objective reality, which can never be known by a single mind. A being's subjective reality can be at least partly reconstructed by field observation and experiment, trying to find out what it perceives by what it reacts to and how. Cooperation of ethicists with ethologists could contribute significantly to identifying and understanding morally sensitive interactions of humans and animals.[20]

Although such approaches may yield a fair approximation, Jakob von Uexkuell and modern ethologists warn us that it would be a delusion to believe that they could lead to being able to perfectly see the world with another being's eyes. Knowledge is limited by the fact that the scientist "is always dealing with events that take place in *his* space and in *his* time and with *his* qualities" (J. von Uexkuell 1926). Therefore, who is being regarded as a moral patient and in what way depends on who the moral agent is. This does not mean that knowledge is not possible, but it reminds us that it is positioned. Reflecting on the role we play in another being's environment, therefore, also provides us with a relationship that is the basis for moral motivation, as well as the possibility to gain the knowledge to act accordingly. Instead of focusing exclusively on the moral patient, we must turn our attention once more to the moral agents, to ourselves.

The Empathic Moral Agent

I suggest a notion of *empathy* that denotes the attempt to reconstruct another's *Umwelt* in a nonjudgmental manner in the awareness of our limited abilities to reach a perfect understanding. This is a rational endeavor, although it does comprise emotional as well as cognitive aspects, which can serve to counterbalance, correct, or reinforce each other in a reflective way.[21] For instance, whereas the cognitive element may help us to be empathic with beings that do not have a positive emotional appeal to us, the emotional element may kindle our curiosity to catch a glimpse of other subjective worlds. Also, moral behavior that is curtailed of any sort of emotional response may be seen as deficient (see Nelson 1997, vii). The balance between cognitive and emotional elements may certainly differ between individual moral agents.

Empathy should not be confused with a number of related terms. First, it is neither some kind of vague intuition nor a sporadic, volatile feeling. Second, it is not identical with sympathy or compassion, which Schopenhauer ([1840] 1988) and Ursula Wolf (1990) suggest as a basis for our moral treatment of animals. In contrast, empathy does not necessarily involve cosuffering as a response to acknowledging another being's reality. Although such a response may happen as an intuitive reaction, it can be balanced by cognitive reflection of the matter. Moreover, the concept of empathy proposed here is not bound to our traditional understanding of suffering at all.[22] Third, empathy is not identical with caring, which usually implies that the person has taken on the matter as a personal responsibility and intends some action.

The form of empathy I am putting forth, which does not rely on personal sympathy or emotional attachments, can be conceptualized as a basic moral duty or rule. Such a duty to empathy can be grounded in the need to justify harmful or manipulative intrusions into other beings' lives. This justification requires an exploration of whom you hurt or wrong and how, in what relationship you are with this being, and what alternatives exist.

Such an empathy-oriented approach can be of relevance to (bio)ethics in several respects:

1. On the conceptual level, empathy is a prerequisite for justice and for care, which represent the two empirically validated moral orientations (Gilligan 1982; Kohlberg 1981) as well as the basic constituents of most ethical theories. Without trying to understand other beings' subjective worlds, one cannot possibly address a "concrete other" (Benhabib 1987) in one's moral reflection and action.

2. On the methodological level, empathy is necessary to discern moral problems. Suffering and abuse may be easily overlooked. In this way, the "faculty of empathy . . . [is] indispensable in providing us with an access to the domain of the moral" (Vetlesen 1994, 6) and thus determines what we perceive as moral problems. It also is a central element in the continuous circle of the search for our identity and moral action, of reflection and proflection.[23]

3. An empathic perspective is also a source of moral motivation; it helps create relationships and thus intensifies moral concern and obligations. By emphasizing the connectedness of beings instead of viewing them predominantly as autonomous individuals, an empathy-oriented approach can capture an important facet of our everyday experience.

Focusing on empathy can help avoid a priori ascriptions of value and also fosters a more comprehensive understanding of concrete moral situations. Who am I, the moral agent, and how do my intuitive response and the cognitive analysis relate in this case? Who is the moral patient, and in what kinds of relationships are we? Who else participates in this moral situation? What are possible and available options for action? Acknowledging the existence of different subjective realities, refraining from the attribution of moral labels with objective claims, and trying to find compromises between one's own and others' interests will foster tolerance and serve as a sensible framework for discourses across cultures or value systems.

Empathy by itself certainly does not constitute an ethic, nor does it replace principles such as justice. But drawing attention once more to its fundamental importance can help to shift the framing of some questions that are particularly relevant in animal ethics. Instead of asking, "Who is worthy of protection?" (e.g., as indicated by quality of life), the more salient question appears to be, "Who is in need of protection?" That way, a ranking of vulnerabilities, and thus situated moral decisions, would gain precedence over the universal ascription of value.

Weighing Options without Labeling

My refusal to accept the strategy of determining someone's moral worth by asking who he, she, or it is and what abilities he has and my preference not to exclude any living being prima facie from moral considerability leave me with the need to suggest an alternative approach to arriving at concrete moral decisions. Prioritizing in specific situations clearly cannot be avoided; the question of whose needs will be met first, or how they are to be balanced against one's own interests, will always remain. At the same time, I am not arguing for a crude egalitarianism. It is important to acknowledge that we do favor some beings over others, our dog at home over a rat, the rat perhaps over a butterfly, the butterfly over a mosquito. But how can we orient ourselves in moral conflicts without being guided by a clear moral ranking of the beings surrounding us?

Again, we cannot avoid implicitly ranking other beings by deciding whom to favor in the case of a conflict of interests. But these decisions should be bound to specific situations determined by the array of participants and available choices. Those decisions are thus of a purely pragmatic character and do not constitute a genuine source of value; de facto they may reflect somebody's current preferences, but those are not taken as normative. The fact that we are living at the expense of others is seen as a limitation for which we have the responsibility to compensate. This compensation can occur by reflecting the fact of our living at the expense of others and by sincerely attempting to find a balance between our well-being and that of others. In this sense, I would agree that a conception of the "good life" seems to be indispensable to ethics (Pauer-Studer 1996) and that normativity is grounded in our sense of identity (Korsgaard 1996)—individually and as a species.

Examining the relationship we have with the moral patient can provide some orientation as to how to weigh a potential intrusion or conflicting interests.[24] For the environment of some beings, we appear only as some unspecified external factor, for example, as the source of water for a plant in our room. In this case, one could pose a negative duty not to unnecessarily interfere with this being's well-being, if it was concluded that this being could in fact be harmed or misused. With other beings we are in a relationship that implies responsibilities to care for this being unless there are good reasons not to. Finally, some relationships with beings are central to our own identity, such as being your father's daughter or being human. These special relationships give extra weight to the responsibilities they constitute. Being a member of the human species has traditionally been part of this kind of relationship, and this is the realm of traditional personal ethics.[25]

How we treat another being in a specific situation therefore does not depend on a general moral status it is being ascribed, its abilities, its similarity to humans, or its contribution to overall happiness. Rather, the decision is influenced by the moral agent's situation and by an array of pragmatic criteria.[26] The way criteria are employed for weighing the need for moral attention thus fundamentally differs from the use of criteria for classificatory purposes of a being's moral worth.

It differs first in the role the moral agent finds itself in. It is not the role of an impartial arbiter applying rules to determine an individual's status. Rather, the moral agent sees her- or himself as a positioned being with certain preferences who is forced to decide among conflicting interests. The agent will therefore not claim objectivity in making the choice but can still defend the decision as a sound or justified one given the circumstances. This means that those decisions cannot be applied to other situations but must be constantly revised and reflected. In their search for the right balance between own and other interests, moral agents will try to change the situation according to the requirements to empathize and minimize instrumentalization of other beings. One cannot simply use a criterion such as the ability to have plans for the future, for example, as a tool to classify beings into "may kill painlessly" and "may not kill painlessly." Such a criterion would serve as a descriptor of how the moral agent perceives the other being's subjective reality and as an indicator of its vulnerabilities, needs, and potential for enrichment. What is perceived as a vulnerability or a need depends on the situational, personal, social, and cultural context. One cannot determine abstractly how to weigh criteria; it must be part of an ongoing moral discourse.

Theoretical and Practical Implications of an Empathy-Oriented Approach

Although justice and fairness considerations have been seen traditionally as guiding our mature moral decisions (Kohlberg 1981), studies in moral psychology point to care as another perspective (Gilligan 1982), whereby this "ethics of care" is claimed to be distinct from and incompatible with an "ethics of justice." Although both sexes can adopt both perspectives, the care perspective is more prevalent among women, whereas the justice perspective has been found more frequently among men. The care perspective focuses not on autonomous individuals but on relationships and responsibilities. It denies the possibility of an impartial moral standpoint and sees moral judgments as dependent on situative or relational contexts.[27]

Gilligan captures the essence of the difference as follows: "The shift in moral perspective is manifest by a change in the moral question from 'What is just?' to 'How to respond?'" (1987, 23). For the empathy-oriented approach I advocate, I would elaborate the second question as "How shall I (or we) respond *in this situation?*" emphasizing the importance of the moral agent's moral identity and the situational context.

Not surprisingly, this presentation of a new perspective of ethics has not remained unchallenged. Besides methodological criticism concerning the empirical data and their interpretation (e.g., Broughton 1983; Nunner-Winkler 1984), the notion of a distinct "ethic" has been the focus of scrutiny. Some critics have argued that the care perspective is actually the application of justice to concrete circumstances (e.g., Held 1995). Others (e.g., Nelson 1992) have wondered if having one universal ethics for all is not a counterproductive construct, particularly for feminist concerns. Should there be different morals for public and private affairs? One for household issues and one for the office? One for men and one for women?

Empathy is a prerequisite of care as well as of justice. Without a concept about how a being experiences the world and who plays a part in it, how could anyone care for it or try to find out what it is due? The empathy-oriented approach therefore cannot be positioned on either side of the "justice-care divide." Nor do I believe that such a divide is adequate for normative ethics. Although different styles in moral argumentation may be interesting phenomena and important to realize, it is a classic insight that ethical reflection needs to encompass care or benevolence as well as justice considerations (see Menzer 1924). Nevertheless, my approach certainly contains some elements that are of concern to feminist ethics (see Lebacqz 1995), among which are a focus on a situated, relational self and on contextual particularities and a suspicion against a priori abstract determinations of moral value presented as fixed hierarchies. Although it is interesting to note that women have been found to be more concerned about the suffering of research animals than men and tend toward a more restrictive position regarding the use of animals in research (see Eldridge and Gluck 1996), the central concern of my approach is neither a claim to feminine expertise in such matters nor the promotion of feminist interests. On the contrary, the main thrust of my argument to refocus the attention on empathy is to demonstrate its relevance as a basic and central concept.

No View from Nowhere

The empathy-oriented approach can avoid some of the pitfalls of strategies, such as Frey's quality-of-life view on the value of life, that rely on an apodictic ascription of moral value. Their claims to objectivity, universality, and impartiality are especially problematic. The empathy-oriented approach tries to avoid these problems by looking more closely at our limitations as human moral agents and by reducing those claims from the outset.

1. *Positioned objectivity instead of pseudo-objectivity.* That our knowledge is imperfect and subject to constant change is rather obvious and is certainly true with regard to our understanding of animals. As Bernard Rollin notes, "[W]hat we consider the nature of a given

animal (or for that matter, of man) will depend in part on the scientific theories and conceptual schemes (biological, ethological, genetic) current at a certain time and place, which theories involve elements of what is traditionally called conventional" (1981, 57). The consequence of such uncertainty is the need for maximal inclusivity concerning our scope of moral concern. I concur with Rollin's conclusion *"that any living thing with interests is an end in itself,* worthy of moral consideration merely in virtue of its being alive. That in turn means that even if we use another living creature as a means, it must never be *merely* as a means, but we should always keep in mind a respect for its end, that is, its life, and the interests and needs associated with that life" (51).[28]

Such a reduction in our epistemological claims need not lead to relativism. Rather, as Hilary Putnam argues, we should "accept the position we are fated to occupy in any case, the position of beings who cannot have a view of the world that does not reflect our interests and values, but who are, for all that, committed to regarding some views of the world—and, for that matter, some interests and values—as better than others. This may be giving up a certain metaphysical picture of objectivity, but it is not giving up the idea that there are what Dewey called 'objective resolutions of problematic situations'—objective resolutions to problems which are *situated,* that is, in a place, at a time, as opposed to an 'absolute' answer to 'perspective-independent' questions. And that is objectivity enough" (Putnam 1993, 156).

2. *Context sensitivity instead of pseudo-universality.* Universality in the determination of a being's moral value can only be claimed if the moral agent is neglected in her or his situational context, which is constituted by personal relationships, cultural background, choices and obligations, and so on. Whereas ascribing value to other beings a priori may be hard to justify, some sort of valuing implicitly occurs in each moral decision. But this attribution of value is always tied to a certain situation and cannot simply be generalized. This means that the situational context gains importance in an empathy-oriented approach. The attention shifts from the attempt to identify a being empirically in order to determine its moral status to the following questions: Where do my actions interfere with other interests, and is there a genuine conflict of interests? How can I minimize that interference,

or can I avoid it altogether? What could a compromise look like? At this point, empirical factors come into play, but secondarily. They do not by themselves confer moral value but instead help to identify where interests are being interfered with.

These considerations will eventually lead to a ranking of the degree of moral attentiveness toward different animals—the treatment of a dolphin might give us more headaches than that of a beetle—but the use of empirical criteria here should be seen as a pragmatic working tool in need of constant revision, with no claims of determining, let alone determining objectively, a being's worth. For example, not the fact that I am sentient confers moral value to me, but the fact that I as a living entity can feel pain means that any infringement on my "interest" not to be hurt must be justified. Both "moral agent" and "moral patient" are woven into this situational context, which shapes the options for action from which the moral agent can choose. Again, renouncing universality in value attribution does not lead to moral indifference but obligates the moral agent to weigh arguments and options anew in each situation.

3. *Balanced reflective partiality instead of pseudo-impartiality.* Although I do not mean to condone either egotism or speciesism, from an empathy-oriented perspective it is not helpful to deny our tendencies to favor ourselves, our family, and our species. As Onora O'Neill (1996) has pointed out, care and concern are necessarily selective. We do have emotions and attitudes as well as a sense of special obligations constituted by relationships of different natures (e.g., kinship, reciprocity, or using a being for one's own purposes), which influence our decision-making process. Unless we acknowledge the partiality of our decisions, we will not be able to reflect on them and eventually counterbalance them. An unreflected preference for our fellow human beings to the disadvantage of nonhuman beings would rightly be criticized as speciesism. But an epistemological anthropocentrism is unavoidable. Our knowledge is limited and positioned by our membership in the human species. Ethical anthropocentrism, on the other hand, can be justified only if we make a conscious, well-reasoned decision for it. This requires ideally that one has empathically explored the subjective realities of all beings that are part of the respective situation. Omitting this step would lead to morally unacceptable

speciesism. What degree of anthropocentrism we consider justified is a societal decision that has to be worked out in public discourse.

Rejecting the (speciesist) argument that human interests count more merely because they are human does not necessarily lead to an unbiased point of view. Rather, I maintain that there is no neutral moral perspective and that assuming or pretending neutrality may be more dangerous than recognizing its impossibility. The scarcity of reflection on the role of our positioned standpoint may be because utilitarian considerations have dominated, although not exclusively shaped, the discussion. Focusing on the moral agent rather than on defining the objects of moral consideration may help to identify problematic aspects of the ascription of moral value to human and nonhuman beings.

We should therefore not view our moral preference for the human species, which most of us demonstrate daily in our use of animal products, as an expression of a superior value of human beings but rather as our—reflected—limitation as human moral agents. The ethologist Frans de Waal formulates his view on humans' relationship to animals in the following way: "Personally, I do not feel superior to a butterfly, let alone to a cat or a whale. But who can deny our species the right to construct its moral universe from a human perspective? It will be up to society to decide whether it will continue to support certain kinds of research on certain kinds of animals" (1996, 215). This means that we cannot rest on our current perspective but have the obligation to continue exploring vulnerabilities and needs of the beings around us and to balance them against our needs.

Reshaping the Debate
on Using Nonhuman Animals in Research

What are the features that an empathy-oriented perspective should emphasize in a fierce public moral debate such as the one around the use of animals in research, in which distorted polemics have prevailed on both sides?[29] First, empathy is a basic communicative condition for a successful discourse. That speakers "strive to reach an understanding of each other's positions" (Gluck and Kubacki 1991, 159) is a crucial element in a constructive, consensus-oriented discussion. Avoiding labeling another position (and the persons holding it) as a group of butchers or ignorants assaulting biomedical science (see Pardes and Pincus 1991) could also help to prevent the deplorable shadow fights between opponents and defenders of animal research, which regularly occur within academia and beyond.

Across the parties is a rather broad agreement that unnecessary infliction of suffering in animals is morally wrong, which should provide a sufficient base for a substantial, factual debate. It may thus be helpful to leave abstract, fundamental questions in the background and focus on concrete situations, trying to explore the subjective realities of all beings involved, looking closely at the spectrum of options available to the moral agent, and then trying to find a consensus on how to weigh them. Consider the example of a frog that is kept for research purposes in a laboratory. Instead of determining the moral value of the frog as a basis for deciding how to treat it, the empathic moral agent I advocate would try to find out more about the nature of this frog's *Umwelt* by posing questions such as "What does it perceive?" and "How does it interact with its environment?" Discovering the manifold ways in which this interplay between the frog and its environment can be intruded upon would further motivate this caring, empathic moral agent to provide opportunities for the frog to create a rich perceptive and operational interaction with its environment and to keep the intrusion to a minimum. The agent then would consider the situational context and explore the options: Is this experiment really needed? Can it be replaced? Could the research be designed differently in the future?

A number of examples already exist for the implementation of an empathy-oriented perspective into the debate, which is captured, for instance, in the famous "Three Rs" of animal research—reduction, refinement, and replacement (see Goldberg, Zurlo, and Rudacille 1996). One such example is the existence of IACUCs (Institutional Animal Care and Use Committees) in the United States and Canada, and of Animal Protection Committees in many other countries, which deal with impending moral problems in their situational context. Another example is the inclusion of a "psychological well-being" clause for primates kept in research facilities as one of the criteria for the IACUC review process (see Mukerjee 1997), which encourages researchers to consider what an environment might look like from the animal's point of view. Finally, there is a noticeable focus on finding alternatives to the use of animals within and beyond universities, in the form of prizes, scholarships, funding, and public education. The systematic incorporation of initiatives such as these would be an important contribution toward a further improvement of the debate on the use of nonhuman animals in research—leading it away from the ascription of moral values understood as objective givens and toward a focus on the process of moral conflict-solving and decision-making based on the striving for empathic understanding of all beings involved.

Conclusion

Are "moral price tags" just shorthand for underlying moral considerations? Is any criticism of the ascription of moral value to beings just linguistic pedantry? I have tried to show that unfounded and potentially dangerous claims to objectivity, impartiality, and universality accompany the determination of moral value. There is an important difference between the apodictic definition of a being's worth and the exploration and reflection of an individual or societal decision, resulting from the search for a balance between one's own and others' interests and which aims at positioned objectivity, contextual sensitivity, and a balanced, reflected partiality.

An empathy-oriented approach refocuses the moral agent's attention on other beings' vulnerabilities and needs and his or her relationship to them. It allows for an integration of nonhuman animals into ethics that does not inherently endanger the moral status of human beings. Such an approach would allow us to avoid resorting to pseudo-objective moral labels and to acknowledge that our decisions are in fact based on our limited knowledge, on the way we experience our world, and on our attempts as individuals and as a species to reflect and balance our preferences.

We often try to draw moral lines in order to be more clear and more secure in our judgment; I believe this is neither a promising nor an appropriate attempt. Instead, we should try to settle for moral minimal standards, as they often are already crystallized in regulations or the law, and then formulate a strong obligation for further consideration of how conflicts can be reduced and moral ideals promoted according to the possibilities warranted by the respective situation—considerations that may eventually change those minimal standards. It is true that in this way we will never know exactly if we have found the optimal solution. But even a constant struggle and uncertainty will be more satisfying than a false safety and moral self-complacency.

Notes

1. In *Introduction to the Principles of Morals and Legislation* (1789), one of the fundamental works of classical utilitarianism, Jeremy Bentham sought to extend the realm of the moral beyond rational (human) beings: "The question is not Can they reason? nor, Can they talk? but, Can they suffer?" (quoted in Singer 1990, 7).

2. For surveys of the extensive discussion, see DeGrazia (1996, 1–9), Regan (1995), Callicott (1995), and Tannenbaum and Rowan (1995).

3. I use the term *speciesism* to denote an unjustified (and mostly unreflected) attribution of superior value to one's own species, as opposed to *anthropocentrism*, which indicates a substantiated and articulated preference for the human species—an argument that cannot be easily dismissed from philosophical debate.

4. Such an argument can be found whenever there is some confusion about the moral status of a being, for example, in the case of human embryos or the brain dead.

5. I propose neither an intuitionist nor an emotivist position here. Rather, I understand empathy as a reflective phenomenon that integrates cognitive and emotional abilities.

6. By *meta-ethical,* I refer to an analysis of argumentations in ethics in a general sense and not to a particular school of thought in British philosophy.

7. This is unlike accounts that attribute intrinsic value to nonhuman beings, such as Regan's (1983).

8. Frey's argument is Kantian in form, if not in content. As is well known, Kant's ethic included only indirect duties in regard to animals, a restriction that of course need not apply to any deontological ethical theory.

9. For example, see Bernard Williams's critique of utilitarianism (1972).

10. Using quality of life as a criterion for deciding who is to benefit from research and who is to be sacrificed could change the notion of provivisectionism in a peculiar way. If some nonhuman primates, for example, have a quality of life that is considered higher than that of a comparable number of humans, wouldn't there be not only a justification but a requirement to use those humans in research for the primates' well-being, given that overall species-specific standards cannot be used in favor of a single individual from a nonspeciesist perspective? The traditional provivisectionist assertion that "We may use animals in research for human benefit" thus turns into "Beings with what is judged to be a lower quality of life may be used for the benefit of those that are regarded as having a higher quality of life."

11. According to Frey, "the value of a life is a function of its quality; its quality is a function of its richness of content; and its richness of content is a function of its capacities and scope for enrichment" (1996, 203).

12. Such criteria, for example, could include species membership, moral agency, or rationality.

13. There is a considerable variety of interpretations of suffering presupposing different abilities. For example, suffering can be understood as a threat to one's integrity as a person (Cassell 1991), as an inherently interpersonal or social phenomenon (Kleinman, Das, and Lock 1997), or as pain or an unpleasant sensory experience (Singer 1979). Arguably, suffering in a more figurative sense need not even involve consciousness. For example, from a nonconsequentialist perspective, a being could suffer an injustice—be wronged—without being aware of it (if someone cheated you and you didn't notice it until much later, you would still likely insist that you had been wronged when the cheating occurred). It is thus a matter of interpretation, whether personhood, communicative abilities, sentience, or awareness are considered requirements for suffering. This question, however, is too closely linked to central philosophical issues of the meaning of existence to likely ever yield a consensus.

14. I do not discuss here the utility of quality-of-life concepts as descriptive instruments, for example, for individualizing therapeutic decisions in clinical medicine.

15. Some authors attribute inherent value only to individual beings, either to those that are "subjects-of-a-life" (e.g., Regan 1983) or to all living beings (e.g., Taylor 1986), whereas others believe that species and ecosystems as such have noninstrumental moral significance as well (e.g., Johnson 1991).

16. I use the term *moral patient* instead of *moral object* to avoid an association with a lifeless, passive thing.

17. I use *autonomous* here as referring to "the capacity of the will to legislate itself" (Allison 1995, 437).

18. Life on the vegetative level (i.e., plants and unicellular organisms) follows a simple control-system model; "self, nonself and nonexistence are still largely undifferentiated on this level" (T. von Uexkuell 1997, 20). Only animal life (containing a nervous system) is able to form a "subjective universe," an *Umwelt* (21).

19. "It" originally denotes the "will-to-live" in Schweitzer's quotation.

20. For an example of ethologists' openness to such an interdisciplinary endeavor, see Bekoff (1994).

21. As recent neurophysiological research has demonstrated, rationality can only be adequately understood as a phenomenon integrating cognition and emotions (Damasio 1995).

22. Empathy may thus be more adequately called "pro-flection," meaning "thinking from the perspective of the other," as opposed to "re-flection," which reconnects those thoughts with the own identity. The German philosopher Franz Fischer (1965) coined this term. However, I do not use any other elements of Fischer's philosophy.

23. The connection between identity and morality is an important topos in philosophy. The philosopher Christine Korsgaard, for example, describes their connection in the following way: "I believe that the answer [to the normative question] must appeal, in a deep way, to our sense of who we are, to our sense of our identity" (1996, 17). Our identity is closely related to how we perceive others and how we think we are perceived by them. Or, as George H. Mead has expressed in his Fragments on Ethics, "We are what we are through our relationship to others" (1967, 379).

24. Relationship is only one aspect in the determination of our moral obligations toward another being. For a subtle account of the multiple criteria that can come into play, see Warren (1997).

25. Mary Midgley (1983), among others, has emphasized the importance of social bondedness in moral argumentation.

26. The fact that all living beings do not need to justify their existence means that humans don't either. This means that intrusions or instrumentalizations that are deemed unavoidable for preserving our existence need no further justification. It is necessary, however, to try to develop less-intrusive alternatives.

27. For succinct summaries of the ethics of care, see Carse (1991) and Sharpe (1992).

28. What counts as a "living being with interests" has to be the object of a discourse; for my position see the previous section, "Rich Life or Not: Who Cares?"

29. For a vivid description of this debate, see Blum (1994).

References

Allison, H. E. 1995. Kant, Immanuel. In *The Oxford companion to philosophy*, edited by T. Honderich, 435–38. Oxford: Oxford University Press.

Bekoff, M. 1994. Cognitive ethology and the treatment of non-human animals: How matters of mind inform matters of welfare. *Animal Welfare* 3:75–96.

Benhabib, S. 1987. The generalized and the concrete other: The Kohlberg-Gilligan controversy and moral theory. In *Women and moral theory*, edited by E. F. Kittay and D. T. Meyers, 154–77. Totowa, N.J.: Rowman and Littlefield.

Bentham, J. 1789. *Introduction to the principles of morals and legislation*. London.

Blum, D. 1994. *The monkey wars*. New York: Oxford University Press.

Broughton, J. M. 1983. Women's rationality and men's virtues. *Social Research* 50:597–642.

Callicott, B. J. 1995. Environmental ethics: Overview. In *Encyclopedia of bioethics*, edited by W. T. Reich, 676–87. New York: Simon and Schuster Macmillan.

Carse, A. L. 1991. The "voice of care": Implications for bioethical education. *Journal of Medical Philosophy* 16(1):5–28.

Cassell, E. J. 1991. *The nature of suffering and the goals of medicine.* New York: Oxford University Press.

Damasio, A. R. 1995. *Descartes' error: Emotion, reason, and the human brain.* New York: Avon Books.

DeGrazia, D. 1996. *Taking animals seriously: Mental life and moral status.* New York: Cambridge University Press.

de Waal, F. B. M. 1996. *Good natured: The origins of right and wrong in humans and other animals.* Cambridge, Mass.: Harvard University Press.

Dworkin, R. 1994. *Life's dominion: An argument about abortion, euthanasia, and individual freedom.* New York: Vintage Books.

Eldridge, J. J., and J. P. Gluck. 1996. Gender differences in attitudes toward animal research. *Ethics and Behavior* 6(3):239–56.

Feinberg, J. 1974. The rights of animals and unborn generations. In *Philosophy and environmental crisis,* edited by W. T. Blackstone, 43–68. Athens: University of Georgia Press.

Fischer, F. 1965. *Proflexion und Reflexion: Philosophische Übungen zur Eingewöhnung in der von sich reinen Gesellschaft.* Kastellaun: Henn.

Fox, W. 1993. What does the recognition of intrinsic value entail? *Trumpeter* 10(3):101.

Frey, R. G. 1988. Moral standing, the value of lives, and speciesism. *Between the Species* 4:191–201.

———. 1996. Medicine, animal experimentation, and the moral problem of unfortunate humans. In *Scientific innovation, philosophy, and public policy,* edited by E. F. Paul, F. D. Miller Jr., and J. Paul, 181–211. New York: Cambridge University Press.

Gilligan, C. 1982. *In a different voice: Psychological theory and women's development.* Cambridge, Mass.: Harvard University Press.

———. 1987. Moral orientation and moral development. In *Women and moral theory,* edited by E. F. Kittay and D. T. Meyers, 19–33. Totowa, N.J.: Rowman and Littlefield.

Gluck, J. P., and S. R. Kubacki. 1991. Animals in biomedical research: The undermining effect of the rhetoric of the besieged. *Ethics and Behavior* 1(3):157–73.

Goldberg, A. M., J. Zurlo, and D. Rudacille. 1996. The three Rs and biomedical research. *Science* 272:1403.

Goodpaster, K. E. 1978. On being morally considerable. *Journal of Philosophy* 75:308–25.

Hargrove, E. C., ed. 1992. *The animal rights/environmental ethics debate.* Albany: State University of New York Press.

Held, V., ed. 1995. Justice and care. In *Essential readings in feminist ethics.* Boulder, Colo.: Harper Collins.

Johnson, L. E. 1991. *A morally deep world.* Cambridge: Cambridge University Press.

Kant, I. [1785] 1994. *Grundlegung zur Metaphysik der Sitten.* Reprint, Hamburg: Meiner.

Kleinman, A., V. Das, and M. Lock, eds. 1997. *Social suffering.* Berkeley: University of California Press.

Kohlberg, L. 1981. *The psychology of moral development: The nature and validity of moral stages.* San Francisco: Harper and Row.

Korsgaard, C. M. 1996. *The sources of normativity.* Cambridge: Cambridge University Press.

Lebacqz, K. 1995. Feminism. In *Encyclopedia of bioethics,* edited by W. T. Reich, 808–18. New York: Simon and Schuster Macmillan.

Mead, G. H. 1967. *Mind, self, and society.* Chicago: University of Chicago Press.

Menzer, P., ed. 1924. *Eine Vorlesung Kants über Ethik.* Berlin: Pan Verlag Rolf Heise.

Midgley, M. 1983. *Animals and why they matter.* Athens: University of Georgia Press.

Mukerjee, M. 1997. Trends in animal research. *Scientific American* 2:70–74.

Nelson, H. L. 1992. Against caring. *Journal of Clinical Ethics* 3(1):8–15.

———. 1997. Introduction. In *Stories and their limits: Narrative approaches to bioethics,* edited by H. L. Nelson. New York and London: Routledge.

Nunner-Winkler, G. 1984. Two moralities? A critical discussion of an ethic of care and responsibility versus an ethic of rights and justice. In *Moral development,* edited by W. M. Kurtines and J. L. Gewirtz, 348–61. New York: Wiley.

O'Neill, O. 1996. *Towards justice and virtue.* Cambridge: Cambridge University Press.

Pardes, H. W. A., and H. A. Pincus. 1991. Physicians and the animal-rights movement. *New England Journal of Medicine* 324(23):1640–43.

Pauer-Studer, H. 1996. *Das Andere der Gerechtigkeit: Moraltheorie im Kontext der Geschlechterdifferenz.* Berlin: Akademie-Verlag.

Putnam, H. 1993. Objectivity and the science-ethics distinction. In *The quality of life,* edited by M. Nussbaum and A. Sen, 143–57. Oxford: Clarendon.

Regan, T. 1983. *The case for animal rights.* Berkeley: University of California Press.

———. 1995. Ethical perspectives on the treatment and status of animals. In *Encyclopedia of bioethics,* edited by W. T. Reich, 158–71. New York: Simon and Schuster Macmillan.

Regan, T., and P. Singer, eds. 1989. *Animal rights and human obligations.* Englewood Cliffs, N.J.: Prentice Hall.

Rollin, B. E. 1981. *Animal rights and human morality.* Buffalo, N.Y.: Prometheus.

Schopenhauer, A. [1840] 1988. Preisschrift über die Grundlage der Moral. Reprint: In *Kleinere Schriften,* 459–632. Zürich: Haffmans.

Schweitzer, A. 1989. The ethic of reverence for life. Translated by J. Naish. In *Animal rights and human obligations,* edited by T. Regan and P. Singer, 32–37. Englewood Cliffs, N.J.: Prentice Hall.

Sharpe, V. A. 1992. Justice and care: The implications of the Kohlberg-Gilligan debate for medical ethics. *Theoretical Medicine* 13:295–318.

Singer, P. 1979. *Practical ethics.* Cambridge: Cambridge University Press.

———. 1990. *Animal liberation: A new ethics for our treatment of animals.* 2nd ed. New York: New York Review.

Singer, P. 1995a. Animal research: Philosophical issues. In *Encyclopedia of bioethics,* edited by W. T. Reich, 147–53. New York: Simon and Schuster Macmillan.

———. 1995b. Animals. In *The Oxford companion to philosophy,* edited by T. Honderich, 35–36. Oxford: Oxford University Press.

Tannenbaum, J., and A. N. Rowan. 1995. Rethinking the morality of animal research. *Hastings Center Report* 10:32–42.

Taylor, P. W. 1986. *Respect for nature: A theory of environmental ethics.* Princeton, N.J.: Princeton University Press.

Vetlesen, A. J. 1994. *Perception, empathy, and judgment: An inquiry into the preconditions of moral performance.* University Park: Pennsylvania State University.

von Uexkuell, J. 1926. *Theoretical biology.* London: Keagan Paul.

von Uexkuell, T., ed. 1997. *Psychosomatic medicine.* Munich: Urban and Schwarzenberg.

Warren, M. A. 1997. *Moral status: Obligations to persons and other living things.* Oxford: Clarendon Press.

Williams, B. 1972. *Morality: An introduction to ethics.* New York: Harper and Row.

Wolf, J. C. 1992. *Tierethik: Neue Perspektiven für Menschen und Tiere.* Freiburg: Paulusverlag.

Wolf, U. 1990. *Das Tier in der Moral.* Frankfurt: Vittorio Klostermann.

The Rhetorics of Animal Rights

Anita Guerrini

Abstract: *In this chapter, Anita Guerrini, a science historian, criticizes the way in which the ethical debate over the use of animals in research has been carried out. She notes that the debate has been hindered from its inception by a strategy whereby the advocates of the various positions attempt to discredit the opposition in ways that deflect attention from the central grounding of their arguments. She categorizes these rhetorical approaches as fundamentally a search for scapegoats and demons rather than a deepened sense of understanding. For example, she argues that in their zest to indict the foundation of the experimental method, many prominent animal advocates have blatantly misrepresented the position of no less a figure than the philosopher René Descartes with (erroneous) claims that he denied that animals could experience pain and distress. She puts forth evidence showing that while Descartes certainly doubted the rationality of animals, he did not doubt that they could experience pain and distress. Her thesis challenges the scholarship of writers such as Peter Singer and Tom Regan. If she is correct, these types of misrepresentations may well have reduced attention to the ways in which individuals knowledgeable about the pain perception of animals could discard or ignore its ethical relevance. This goes to the heart of the controversy concerning the moral standing and treatment of animals in research.*

The remaining parts of the chapter outline the extent to which religious language, orientations, and biases have found their way into the debate. In a provocative revelation, Guerrini traces some of the antivivisectionist fervor in nineteenth- and twentieth-century Germany, and to some extent England, to a hatred of the Jews. In so doing, she makes a historical connection between kosher practices, the prevalence of Jewish doctors, their involvement in the research profession, and the antivivisection movement. Guerrini also argues that the polar sides of the debate assert the correctness of their views by invoking the language of religion to support their perspectives. In this sense, animal advocates denote attitude shifts by individuals that move in their favor as "conversion" experiences,

while animal-use advocates have adopted the Judeo-Christian concept of a moral
community to support their exclusion of animals from full ethical consideration.

While the title of this essay is "The Rhetorics of Animal Rights," I use
both "rhetorics" and "animal rights" rather loosely; I am not a rhetori-
cian, nor a philosopher, but a historian with some stories to tell. Aristotle
defined rhetoric as the observation of the available means of persuasion,
and in the cases I will discuss, persuasion is the goal. I am particularly concerned with how people use language in particular circumstances. My
theme is the demonization of the other: how opponents view each other
as existing outside their own moral sphere.

Although I will talk about animal rights as a specific philosophical position, the "animal rights" of the title is simply a shorthand for various pro-
animal views. For example, neither the seventeenth-century opponents of
Descartes nor the nineteenth-century antivivisectionists believed that animals had rights. The modern animal-protection movement includes utili-
tarians such as Peter Singer who do not argue from a point of view of
rights. I do not wish to add to existing fuzzy language, and in the text that
follows I will use the term "animal rights" only in its specific modern sense.

This essay consists of three case studies. In the first example, I will
look at how René Descartes has been demonized to become the villain
of Western science. Is the usual view of Descartes justified? The second
case study concerns the connections between the nineteenth-century antivivisection movement and anti-Semitism in that period. In the final example, I examine some of the "demonizing" language used by those who
oppose or favor animal experimentation today, examining in particular
two popular studies of pro-animal movements.

The Demonization of Descartes

The seventeenth-century philosopher René Descartes is a major rhetorical
target for modern animal activists. In 1982 a member of the radical Animal
Liberation Front slashed a portrait of Descartes at the Royal Society in London. The Australian philosopher Peter Singer, author of *Animal Liberation*
(1975, 200), characterizes Descartes's ideas about animals as the "absolute
nadir" of Western thought on that topic. In *The Case for Animal Rights* (1983),
American philosopher Tom Regan spends an entire chapter attacking

Descartes's views: "[I]t is tempting," he admits, "to dismiss Descartes's position . . . as the product of a madman" (5). In virtually every modern discussion of animal experimentation, Descartes's name arises.

The usual reason cited for Descartes's villainous reputation is given in a report by the nonpartisan Institute for Medical Ethics: "Descartes's denial that animals (despite all appearances to the contrary) were able to suffer, appears to have been widely used as a justification for experimenting on live animals, at a time when that practice was becoming more common" (Smith and Boyd 1991, 300). Yet an examination of contemporary practice, and indeed Descartes's own practice, sheds doubt on this commonly held belief. Few in the seventeenth century or since have believed that animals are not sentient, and almost no one, then or now, has used this notion to justify experimentation. It seems clear that Descartes has been demonized by modern commentators.

In this section, I examine first what Descartes believed about animals and their capacity for sentience and cognition. Second, I look briefly at some contemporary and later experimenters to determine the extent of Descartes's direct influence. Third, I attempt to trace the "demonization" of Descartes back to its eighteenth-century origins.

What did Descartes believe? As the philosopher John Cottingham (1978) has noted, Descartes's statements on animals are by no means clear or consistent. The essence of Cartesian mechanical philosophy was that the world was a collection of mechanisms that could only be looked at and analyzed in mathematical and mechanical terms. By this argument, no distinction existed between what we would call the physical and biological worlds. Everything could be analyzed in terms of the laws of mechanics, including human and animal bodies. In his *Traité de l'homme* (*Treatise on Man,* 1664), Descartes attempted to describe just such a mechanical man.

To Descartes, the body was not what distinguished humans from other animals. On the level of the body, humans and animals were very similar. His revelation of "Cogito, ergo sum" defines the mind as the essence of humanity. Since humans could think, they necessarily had knowledge of God (Descartes's second clear and distinct idea), and therefore they possessed immortal souls. Soul and mind were inextricably intermingled, perhaps even identical. The essence of the world, to Descartes, was this dualism of mind and body, the complete separation of matter and spirit. To many modern commentators, this was the beginning of the end as far as our relationship with nature is concerned: once Descartes severed mind from body, the way was clear for modern science, with its vision of nature as a dead machine and its complete disregard for spiritual matters. But to Descartes, this dualism was a theological as well as a philosophical principle. It guaranteed the primacy of the soul.

Thomas Aquinas, who first proposed that animals were like machines, had been impressed with the new mechanical clocks of the thirteenth century. Similarly, Descartes marveled at the clockwork automata of his time, such as the famous mechanical fountain at Saint Germain-en-Laye outside Paris. If humans could make such convincing devices, how much more skilled was the hand of God, who made the infinitely more complex animal machine? In the *Discours de la méthode* (*Discourse on Method,* [1637] 1968), Descartes asserted that humans could conceivably make a mechanical animal that would be indistinguishable from the real thing. But a mechanical human, however accurate, could never be mistaken for a real human because it would lack a mind, and hence also a soul. Descartes distinguished between the brain and the mind: the mind was spiritual, not material; although the brain could perceive and imagine, it could not reason without the mind. The mechanical human would manifest its inadequacy in two critical respects: it would lack speech, and it would lack the ability to reason.

Descartes believed that animals could neither speak nor reason, and therefore they were simply body, mere machines. The ingenuity of their construction enabled animals to emit sounds in response to certain stimuli or to act in certain ways. As he detailed in the *Treatise on Man,* the body was so formed that it could do quite a lot without reference to the mind. The fact that animals could do things that humans could not was no argument for their possession of reason; clocks, after all, could tell time better than humans could on their own.

Although the body possessed sentience—the ability to feel—this did not imply cognition, the ability to think. In another treatise, *Les passions de l'âme* (*The Passions of the Soul,* [1641] 1971), Descartes stated that our bodies perceive hunger, thirst, and other natural appetites, including the feeling of heat, cold, and pain. The body could feel, but only the mind could think and therefore consciously experience that feeling. The functions of the mind included memory, conscious perception, and most

important, reason. Speech manifested the existence of mind. He explained
this in a well-known passage in the *Discourse:*

> For it is particularly noteworthy, that there are no men so dull-wit-
> ted and stupid, not even imbeciles, who are incapable of arranging
> together different words, and of composing discourse by which to
> make their thoughts understood; and that, on the contrary, there is
> no other animal, however perfect and whatever excellent disposi-
> tions it has at birth, which can do the same. . . . And this shows not
> only that animals have less reason than men, but that they have none
> at all. ([1637] 1968, 74–75)

Descartes ([1649] 1971) acknowledged that animals could be trained
to emit certain sounds, but this always occurred only in the presence of a
certain stimulus, not spontaneously. If animals emitted such sounds, he
added, it was merely to express their feelings. Cottingham views this ad-
mission that animals have such feelings as "fear, hope, or joy" as extraor-
dinary in the context of Descartes's views on animal cognition (1978, 557).
Yet the view that the passions are essentially irrational was not new but a
commonplace of Christian theology.[1] Descartes himself contrasted action,
which is a product of will and therefore of the soul, with the passions,
which he considered to be thoughts arising from some particular agitation
of the animal spirits and although felt directly by the soul, not produced
by the soul. Therefore, Descartes believed that animals did feel pain as a
nervous phenomenon, but that they did not experience it cognitively. Did
he then believe that animals could suffer? Historian Martin Pernick has defined suffering as "the emotional effects of pain, as distinguished from its
physical effect on the body" (1985, 295–96n. 4). But is that emotional effect a product of consciousness or instinct? Another historian, Roselyne
Rey, defines Descartes's view thus: "[T]he animal does not suffer because
it does not think that it suffers" (1993, 94).[2] As she points out, this view
had considerable theological backing. Augustine had declared that no suffering was without purpose. Yet what could be the purpose of the suffering of an innocent beast that did not carry the sin of Adam on its soul, a
beast that, indeed, did not possess an immortal soul at all?

It seems clear on this evidence that Descartes did not believe that animals suffered pain in the same way in that humans did. Yet, to return to
Cottingham, Descartes's assignment of feelings such as joy, fear, and pain

to animals undercut the perfect dualism that would make animals wholly machine-like.

Descartes himself experimented little on live animals. In 1639 he described to the Dutch physician Plemp his vivisection of a rabbit, as well as experiments on the hearts of eels and fish. To his friend Mersenne in 1646 he described embryological observations he had made on the developing chick. He also arranged to have killed a cow he knew to be pregnant so that he could observe the embryo, and he received further specimens of unborn calves from butchers. His anatomical writings include accounts of dissection but not of vivisection. In general, experimentation was not central to Descartes's philosophical program. He regarded experimentation not as a method of discovery but as an aid to deduction by mechanical principles. As philosopher Daniel Garber states, "[W]hile experiment might function as an auxiliary to a deduction, it is the deduction itself and not the experiment that yields the knowledge" (1993, 305).

Descartes's assertion that animals were essentially automata immediately brought forth a torrent of criticism from his contemporaries, and until his death in 1650, he spent much ink answering their objections. While several philosophers and theologians defended his claim, few experimenters did (see Rosenfeld 1940, app. B, C, D). The idea that animals were machines (or at least acted as if they were), or the "beast-machine" concept as it came to be dubbed, was far more important to seventeenth-century researchers as a description of an approach to research than as a description of what animals are really like. Most researchers in the seventeenth century followed Descartes to the extent of looking for evidence of mechanism in animal form and function; but they did not necessarily believe, in consequence, that the animals on which they experimented did not suffer or felt no pain.

The supposed inability of animals to suffer was, with a few exceptions, simply not used as a reason for doing animal experimentation in the seventeenth century. However, two well-known cases, involving the Port-Royal monastery and the clergyman Malebranche, document gratuitous cruelty to animals in the name of Descartes, cases that modern critics of Descartes repeatedly cite. In the first case, the Jensenists, a new religious order centered at the Parisian monastery of Port-Royal, strongly supported the Cartesian philosophy. In *La logique ou l'art de penser* (Port-Royal Logic) (Arnauld and Nicole [1662] 1970), the authors use as an example of a conditional syllogism the proposition "Tout sentiment de douleur est une pensée," that is, "all feeling of pain is a thought." This led to a consideration of beasts and their souls. The chapter concludes:

> Nulle matière ne pense:
> Toute âme de bête est matière;
> Donc nulle âme de bête ne pense. (280)
> (No matter thinks; the entire soul of the beast is matter;
> therefore no beast thinks).

Despite Descartes's concern that the Catholic Church find his writings acceptable, the Church officially condemned his works a decade after his death. The Jesuits, Descartes's former mentors, became strongly anti-Cartesian, perhaps with a certain sense of betrayal by a favored student. It is not entirely surprising that the Jansenists, who believed in the Cartesian "beast-machine" concept so far as to act out its consequences, were also fervent opponents of the Jesuits. A secretary to the Jansenist fathers described their cruelty to animals in a much-quoted passage:

> They administered beatings to dogs with perfect indifference, and made fun of those who pitied the creatures as if they had felt pain. They said the animals were clocks; that the cries they emitted when struck, were only the noise of a little spring which had been touched, but that the whole body was without feeling. They nailed poor animals up on boards by their four paws to vivisect them and see the circulation of the blood which was a great subject of conversation. (quoted in Rosenfield 1940, 54)

In the second case, the Cartesian clergyman and philosopher Nicolas Malebranche (1638–1715) also took the "beast-machine" argument literally. He is reported to have kicked a pregnant dog at his feet and to have responded coldly to a protesting observer, "Eh! Quoi, ne savez-vous pas bien que cela ne sent point?" ("What! Don't you know that it can't feel at all?") (Rosenfield 1940, 74). These two cases still make us cringe, but they appear to be the only instances in the seventeenth century in which the beast-machine concept determined the treatment of animals by humans.

Apart from these two cases, I have found no one in this period who explicitly employed Descartes's notion of the "beast-machine" as a justification for experiments. Rather, many experimenters in the seventeenth and eighteenth centuries used the rhetoric of suffering and readily ascribed feeling to animals. It is not surprising that this was the case in England, where Cartesian theory was never fully adopted. The seventeenth-century experimenter Robert Boyle, for example, described a viper subjected to his vacuum pump as "furiously tortured" (1670, 2044), and his colleague Robert Hooke objected to performing an open thorax experiment on a dog "because of the torture of the creature" (Birch 1772, 498).[3]

But non-English experimenters used similar language. The Italian Carlo Fracassati described a dog who had been injected with vitriol: "[T]he Animal complained a great while, and observing the beating of his breast, one might easily judge, the Dog suffered much" (1667, 490). Another Italian, Giuseppe Zambeccari, did not mention that his experimental animals felt pain when he performed abdominal surgery; however, following an operation, he described a dog as "happy and alert" ([1680] 1941, 323). The Danish physiologist Niels Stensen (Nicolaus Steno) complained in 1661 about his work on dogs, "[I]t is not without abhorrence that I torture them with such prolonged pain" (quoted in Ruysch [1665] 1964, 34). He added, "The Cartesians take great pride in the truth of their philosophical system, but I wish they could convince me as thoroughly as they are themselves convinced of the fact that animals have no souls!" (quoted in Ruysch [1665] 1964, 37). As a Catholic convert (later a bishop), Stensen would have been especially aware of Christian debates on animal soul. His Dutch contemporary Frederik Ruysch ([1665] 1964) refused to perform experiments on live animals because of the cruelty involved. In the eighteenth century, to give only two notable examples, the Reverend Stephen Hales (1731) noted fear, pain, and "uneasiness" in the dogs on which he experimented, while the Swiss physician Albrecht von Haller ([1753] 1935) conducted a study whose premise was that animals felt pain; at the beginning of his description of this research, he apologized for his cruelty.

For most scientists, the Cartesian question doesn't come up. Claude Bernard, the great nineteenth-century French physiologist, expressed what is probably the usual view of scientists: philosophical arguments are fine but have nothing to do with science. "No one knows or bothers to know whether Harvey or Haller were spiritualists or materialists; one knows only that they were great physiologists" ([1878] 1974, 32).

By the end of the seventeenth century, Descartes's "beast-machine" doctrine was heavily criticized from a number of points of view. Some

Catholics felt Descartes left too little room for God and led to the slippery slope of materialism, while some atheists felt that he gave humans entirely too privileged a position. The Jesuit Gabriel Daniel, in his 1690 *Voiage du monde de Descartes* (*Voyage to the World of Descartes*), specifically attacked and refuted the doctrine of animal automatism, arguing the Aristotelian position that animals have what was called a "sensitive soul," which allowed for certain sorts of rational behavior. On the other hand, the mid-eighteenth-century *philosophe* La Mettrie criticized Descartes for not being materialist enough; in his *L'homme machine* (*Machine Man,* 1747), La Mettrie attributed all behavior and thought to material causes.[4] Descartes had by no means settled the question, but his doctrines continued to be the touchstone for discussion, even when no one, apparently, believed them. Toward the end of the eighteenth century, the English philosopher Jeremy Bentham took yet another swipe at Descartes in his famous statement, "[T]he question is not, Can they *reason?* nor, Can they *talk?* but, Can they *suffer?*" (1789, 143).

So why are Regan and others so exercised about Descartes? The argument of Regan and other rights theorists is modeled on Descartes's in that it assumes that cognition is the basis for assigning rights. Regan (1983) in fact only slides the gauge a little further down the chain of being in assigning rights to upper mammals but not to others. Rights theorists have since expanded their range, but the basis of their argument remains the same.

To modern critics, Descartes represents in its purest form the soulless, arrogant, uncaring scientist. Certainly Descartes was not lacking in arrogance, and his own statements give plenty of ammunition to his critics, then and now. He was confident that human rationality made him superior to other creatures, and he was, we must admit, smug in his assurance that humans alone have souls and know God. During the Enlightenment, people were already beginning to doubt that God would take care of them, but they believed fervently that reason would ultimately carry humans through their trials and that humankind would inevitably progress to fuller, happier, more rational lives.

In our postmodern disillusionment, we are no longer sure of any of these things. Yet that soulless modern scientist helps make our lives—extraordinarily comfortable and healthy by seventeenth-century standards—possible. Descartes has become the scapegoat for our collective guilt.

The Rhetoric of Anti-Semitism

I now move forward to the nineteenth and early twentieth centuries to examine a different kind of rhetoric, that of anti-Semitism. In an undergraduate class I teach on the history of animal experimentation, one of the assignments is a radio address by Hitler's henchman Hermann Göring on antivivisection. Many of the students see it as an example of good ideas being held by bad people, but there are deeper questions here. The connection between anti-Semitism and antivivisection goes back further than the Nazis, at least to the middle of the nineteenth century. While the "demonization" of the Jews by Christian Europe dates to the dawn of Christianity, new and disturbing rhetoric emerged with the development of modern biological science in the nineteenth century (Carmichael 1992).

The nineteenth-century German philosopher Arthur Schopenhauer, in his *Über die Grundlage der Moral* (*On the Basis of Morality*, [1841] 1965), argued strenuously against Kant's view that cruelty to animals was morally wrong only insofar that it inclined men to be cruel to each other. Kant's view that "man exists as an end in itself" was, he said, a theological, not a philosophical, view. Not only was it illogical; it led to the immoral corollary that "beings devoid of reason (hence animals) are *things* and therefore should be treated merely as *means* that are not at the same time an *end*." "Thus only for practice," he wrote with sarcasm, "are we to have sympathy for animals" (95–96).

The Romantic German biological theory known as *Naturphilosophie*, which opposed a mechanistic view of nature, strongly influenced Schopenhauer. The *Naturphilosophen* argued that nature was essentially one, that mind or reason derived from nature by a developmental process, and that there was no strict Cartesian division between mind and body (see Trohler and Maehle 1987).[5] However, Schopenhauer did not view animals as equal to humans. Like Descartes, he argued that animals could perceive but not reason; they lacked concepts and the ability to formulate abstract thought. But unlike Descartes, Schopenhauer believed that they had a consciousness of themselves. Lack of reason was not a sufficient criterion to omit them from ethical consideration; compassion, he said, was the only true moral motivation. Because animals could suffer, they could be the recipients of compassion.

"The moral incentive advanced by me as the genuine, is further confirmed by the fact that *the animals* are also taken under its protection," wrote Schopenhauer. The "revoltingly crude" idea that we have no duties to animals was "a barbarism of the West, the source of which is to be found in Judaism." "The essential and principal thing" in animals and humans is the same. Only the "Judaized despiser of animals and idolater of the faculty of reason" could believe differently. Schopenhauer believed that Christian morality originated in India and therefore had much in common with Buddhism and Hinduism, which he admired; "but unfortunately it fell on Jewish soil" ([1841] 1965, 177–78). In Judaism Schopenhauer found the origins of the radical division between animals and humans, a hierarchical view of nature he rejected (see Brann 1975).

It is tempting to dismiss Schopenhauer's views as aberrant; but the rhetoric of anti-Semitism remained tied in various degrees to sympathy for animals and particularly to antivivisection in the nineteenth and early twentieth centuries. Many European universities in the eighteenth century had relaxed their religious tests and allowed Jews to enroll. Historian Fritz Ringer (1969) has argued that because Jews continued to be barred from government posts in the nineteenth century, they turned to the "free" professions of medicine, law, and journalism, as well as university teaching. However, Jews found it difficult to advance in the academic ranks. Ironically, their very success in these fields also made them targets of anti-Semitic attacks, including attacks by antivivisectionists.

A new stereotype, that of the Jewish scientist, replaced the medieval stereotype of the Jewish moneylender. But this new stereotype had long roots: in the Middle Ages, Jews were accused of ritual murder of children. The Jewish practice of ritual slaughter of animals for food, in accordance with kosher law, was often pointed to as evidence of Jews' innate cruelty, the implication being that they enjoyed viewing suffering. This had further resonance because the Jews were also viewed as the murderers of Christ, who was symbolically depicted as a lamb. The imagery of ritual slaughter was often connected with vivisection; even today we use the language of "sacrifice," with all its religious connotations. In 1927 a Nazi member of the Reichstag proposed laws banning both vivisection and ritual slaughter, and a law banning kosher slaughter passed in 1933, after Hitler's takeover (Sax and Arluke 1995).[6] One of the most notorious of the Nazi propaganda films, the 1940 "documentary" *Der ewige Jude (The*

Eternal Jew), concluded with a graphic representation of ritual slaughter, which was so revolting that the narrator advised the squeamish to close their eyes (Hull 1969). As the quintessential outsider in Christian Europe, the Jews carried with them the connotation of "polluters" or "defilers," threats to the dominant culture (Douglas 1984). As we shall see, modern animal protectionists also use the language of pollution.

One of the most prominent converts to the antivivisection cause in nineteenth-century Germany was the composer Richard Wagner, who explicitly linked antivivisection and anti-Semitism. In 1879 Wagner announced his support for the antivivisection cause with a letter to Ernst von Weber, vice-president of the Dresden Animal Protection Society and an ardent antivivisectionist. Weber's own pamphlet, *Die Folterkammern der Wissenschaft* (*The Torture-Chambers of Science*), had been published earlier in the year and was a huge success. Wagner's letter to Weber was published in the local newspaper in Bayreuth, the *Beyreuther Blätter,* and Weber distributed additional copies.

The *Beyreuther Blätter* was known as an anti-Semitic organ, and Wagner's letter linked vivisection to the pervasive influence of the Jews on European culture, demonstrated by the fact, he claimed, that most vivisectors were Jews (Sax and Arluke 1995). In the previous year, Wagner had contributed another essay to the *Beyreuther Blätter,* called "What Is German," which argued that Jewish intrusion into society had contaminated the pure German spirit (Katz 1986). His infamous essay "Judaism in Music," published anonymously in 1850 and reprinted with Wagner's name attached in 1869, portrayed Jewish artists as incapable of true artistic expression; music such as Mendelssohn's was merely entertaining. Jews were the "other," the outsiders, who could not share in the German *Volksgeist* that was the basis of true art. Ringer (1969) has pointed out that during the 1870s and 1880s anti-Semitic sentiments were on the rise in politics and in academe.

The concept of "pollution" once more entered, and Wagner interpreted this in increasingly literal ways. He was convinced that German culture was becoming degenerate owing to Jewish influence and that the way to regenerate it was a multipronged process. It included the expulsion of Jews but also a spiritual and physical regeneration of the German *volk,* the Romantically idealized German people. The spiritual, antimaterialistic German race felt a metaphysical unity with animals, which compelled them to compassion. Wagner also linked this regeneration to vegetarianism, but not, as one might think, because of the cruelty issue. His 1881 essay "Heroism and Christianity" outlines the theory of racial degeneration of Joseph Arthur Gobineau, who postulated an original "pure" Teutonic race. Wagner asserted that vegetarianism was the "Ur-diet" of natural man, whose bodily and moral corruption stemmed from the time he began to eat animal blood (Katz 1986; Sax and Arluke 1995). The notion that Adam was a vegetarian was very old, but the theme of pollution, here blood pollution, again emerged, and Wagner again linked it to the pernicious influence of Judaism and the imagery of slaughter. In his operas, according to critic Marc Weiner, Wagner's "heroes are associated with beautiful, lithe, and powerful animals, while those figures evincing traits associated with Jews, such as avarice, egotism, and lovelessness, are likened to lowly, disgusting, and clumsy creatures" (1996, 90–91). Wagner's friend Friedrich Nietzsche described primitive, uncorrupted man as a "blond beast," a beast of prey, a rhetoric that became popular among the Nazis (Sax and Arluke 1995, 233). The incongruity between these carnivorous beasts and Wagner's vegetarianism apparently remained unremarked.

The rhetorical links between Jews and vivisection were not confined to German-speaking countries. The English antivivisection leader Frances Power Cobbe published an appeal in the *Jewish Chronicle* in 1891 to "the well-known humanity of English Jews" to protest against their fellow Jews who were vivisectors. "Throughout Germany and Austria," she claimed, "the great majority of Vivisectors are Jews" (Cobbe 1891). Cobbe appears here to be swallowing whole the rhetoric of Wagner and others about "Jewish" science, with her own peculiarly nationalistic twist. There were, certainly, few Jewish scientists in England. Jews had only been allowed to matriculate at Oxford and Cambridge in the 1850s, although the newer, secular universities such as the University of London had no religious tests.

Cobbe had begun her career as an antivivisectionist by protesting against foreign vivisectors: Moritz Schiff in Florence and the students at the Alfort veterinary school near Paris (Guarnieri 1987). Although she was not slow to condemn English vivisection as well, there is a certain amount of xenophobia in her condemnation of "foreign" science. In the 1820s, some English researchers had similarly condemned François Magendie's work in France (Manuel 1987).

English Jews were quick to respond to Cobbe's appeal. A flurry of letters in the *Jewish Chronicle* noted that many vivisectors were Christian and
condemned the anti-Semitism implicit in Cobbe's remarks. A few years
later, one Morris Rubens wrote a passionate pamphlet, *Anti-Vivisection exposed, including a disclosure of the recent attempt to introduce anti-Semitism
into England.* Cobbe defended herself against charges of anti-Semitism,
pointing to Jews who actively supported her, and several Jewish women
published letters of support in Cobbe's own journal, *The Zoophilist.* Cobbe
had, it was true, published similar appeals to other religious groups, including Catholics and Quakers (French 1976).

Nonetheless, there was an undercurrent of anti-Semitism in Cobbe's
movement that her appeal only made more plain. The antivivisectionist
Charles Adams made much of the Jewishness of Ernest Hart, editor of
the *British Medical Journal,* referring to him as "E. Abraham Hart" and re-
marking on his Jewish blood in at least one published statement (French
1976, 347). In 1881 *The Zoophilist* published a review of the German antivivisectionist Friedrich Zollner's *Über den Wissenschaftlichen Missbrauch der
Vivisektion* (On the Scientific Abuse of Vivisection). The anonymous re-
viewer agreed with Zollner's "tracing many evils to the uncongenial influences of Judaism and Materialism. It would be wrong to say that vivisection is a Jewish pursuit, yet medicine is, in Germany at least, an
eminently Jewish profession" (quoted in Sax and Arluke 1995, 258). In the
1920s, novelist Dorothy Sayers implicitly acknowledged the pervasiveness
of the image of the Jewish vivisector in her novel *Whose Body?* (1923) in
which she neatly reversed the stereotype. Her villain is an English vivisector whose victim is a Jewish banker.

Much more could be said on the subject. Certainly the rhetoric of anti-
Semitism became more pronounced in the Nazi era; yet the fundamentals
of this discourse, and its connection to antivivisection, existed in the nineteenth century. Anti-Semitism, unconscious or not, was widespread in this
era. England did not have an ideology such as Wagner's, but the reactions
of Cobbe and Adams are revealing. The connection of anti-Semitism to
vivisection and to science reveals the essential fear of modern science held
by Wagner and his German comrades. The historian Fritz Stern (1961) has
detailed the connection between anti-Semitism and antimodernity. Wagner and his intellectual circle professed entirely to reject modern medicine,
not only because of its reliance on vivisectional methods but because it

was essentially unnatural; diet and other noninterventionist cures were more suited for a regenerated race. "Jewish materialism" threatened the Romantic, idealized Germany exemplified by Siegfried in Wagner's *Der Ring des Nibelungen*. Despite the presence of such groups as Jews for Animal Rights, anti-Semitism is not entirely absent even from the modern rhetoric of animal rights. In their book *The Animal Rights Crusade* (1992), discussed in the next section, James Jasper and Dorothy Nelkin describe several recent incidents. This is evidence not for the uniqueness of pro-animal rhetoric, but rather evidence of how deeply it is embedded in the broader culture.

Rhetoric in the Modern Debate

In this section, I address some of the rhetoric used on both sides of the modern debate about animal experimentation. Each side has tried to demonize the other by portraying it as outside of mainstream sentiment. Both sides also employ religious imagery. The predominance of women in the antivivisection movement has meant that much of the rhetoric employed is also highly gender specific. I concentrate on the analysis offered by two recent books, Susan Sperling's *Animal Liberators* (1988) and James Jasper and Dorothy Nelkin's *The Animal Rights Crusade* (1992).

Sperling is an anthropologist, trained in primatology. Her book, based on her doctoral research, was highly criticized both for its techniques and for its conclusions. It is an intensely personal book, a combination of autobiography, journalism, and anthropology. Interviews with animal-rights advocates in the Berkeley area comprised one of Sperling's main sources. Her interviewing technique is rather amateurish and unskilled, but the large sections of verbatim quotations that dot her text provide useful source material. Although she ultimately comes down in favor of experimentation, her ambiguity on the topic for much of the book allows her to play to a full range of metaphors on both sides of the issue.

Jasper and Nelkin's book is quite different. The authors are academic sociologists, and their book is strictly academic and seemingly dispassionate. Sperling's doubts and worries about her own position on the issues constantly intrude on her text, but readers know as little about Jasper and Nelkin's personal beliefs at the end of the book as they did at the beginning. I focus on these books as examples of two different approaches.

Both books agree on many of the rhetorical devices used by both sides; in addition, they are revealing in their unspoken assumptions. I focus here on two specific clusters of metaphors and imagery: religious and rational/irrational.

Sperling asserts that even in a post-Christian culture, "Christian cosmological assumptions" remain an essential framework for discourse (1988, 48). She describes this as an assumption of a Covenant or moral community. Within the Covenant, all are brothers; for animals to be moral equals, they must be within the community. This is another expression of the doctrine of us and the other: for animals to be included, they must be us. This differs from the "stewardship" argument that some environmentalists have used, which contends that we have a moral obligation to care for nature as stewards, without assuming moral equality. Sperling acknowledges that fundamentalist Christians would call this notion of moral equality an inversion of the traditional order; but to her this is proof that this common cosmology is still valid.

In their very title, *The Animal Rights Crusade,* Jasper and Nelkin use the language of religious struggle, and in their text they liken the animal-rights movement to a moral crusade. The movement is divided between "fundamentalists" and "pragmatists." Yet they deny Sperling's claim that we still operate within a Christian cosmology, citing moral philosopher Alisdair MacIntyre's (1984) argument that Western societies have lost the ability to ground moral arguments in convincing ways. Without mutually accepted moral principles, they continue, both sides rely on religious metaphors and what MacIntyre calls "retreat to fundamentalism, reifying their own principles as the ultimate Truth" (Jasper and Nelkin 1992, 175). In the absence of a single dominant religion, any strong belief holds equal weight. While moral arguments on issues such as abortion or animal use often rely on religious fundamentalism, other fundamentalisms, such as the belief that American society is inherently racist or sexist, can also be used as a moral focus. Proponents and opponents on both sides of these issues, the authors note, use religious language without necessarily having religious belief. The use of the term *sacrifice,* with its multiple meanings, is one example. Yet one can also use the language of fundamentalism as a way to demonize the opposition as fanatical. "Thus," state Jasper and Nelkin, "protecting animals sometimes inspires the shrill tone, the sense of urgency, and the single-minded obsession of a fundamental crusade." They add, "The fundamentalist belief that they [animal protectionists] possess the Truth makes them quite intolerant" (41).

Some historians and activists view this recognition of the moral status of animals as a reaction against the deemphasis on the spiritual in modern culture, thereby claiming the moral high ground against irreligious scientists. According to clergyman-activist Andrew Linzey (1987), loving animals is "true spirituality." In the 1960s and 1970s, a school of "ecotheology" developed that attempted to reconcile Christian ideals with respect for nature. To others (e.g., Nash 1989, chap. 4), a religion of nature, often based on Eastern or Native American ideas, could replace a fundamentally exploitative Christianity. Jasper and Nelkin note, "Amidst the moral confusions of contemporary culture, animal rightists offer a clear position based on compelling principles" (1992, 175).

In dividing the material from the spiritual, Descartes also separated nature from spirit. As a consequence, modern science, it is argued, is fundamentally antispiritual. Sperling finds parallels between antiexperimentation protests and "philosophies of personal revelation such as evangelical Christianity" (1988, 149). In both, science is viewed as opposed to emotion and revelation. She goes on to discuss "the sense of revelatory vision through animals," which goes beyond a recognition of their spiritual value, and describes some activists to whom the recognition of speciesism came as a revelation that can be compared to a religious conversion experience (150). Yet this language of evangelicalism can also be seen as one way whereby Sperling characterizes animal activists as fanatics, out of the mainstream. Jasper and Nelkin also describe the conversion experiences of activists (1992, 45).

The language of pollution, which was prominent in anti-Semitic arguments, is also prominent among animal rightists. Sperling details the Victorian notion that new technologies and medicines "polluted" the natural human body, and modern activists compare this pollution to environmental degradation (1988, 152–53). The blood in the Christian tradition is associated with healing; contact with blood is a source of pollution in the rhetoric of animal rights. Jasper and Nelkin cite antifur slogans such as "Wearing their fur is as glamorous as drinking their blood" (1992, 152). To a vegetarian, blood is disgusting. Sperling quotes a vivid passage from Peter Singer's *Animal Liberation:*

Vegetarianism brings with it a new relationship to food, plants, and nature. Flesh taints our meals. Disguise it as we may, the fact remains that the centerpiece of our dinner has come to us from the slaughterhouse, dripping blood. Untreated and unrefrigerated, it soon begins to putrefy and stink. When we eat it, it sits heavy in our stomachs, blocking our digestive processes until, days later, we struggle to excrete it. (quoted in Sperling 1988, 153–54)

Both Sperling and Jasper and Nelkin note that animal activists often employ the language of millenarianism, even of apocalypse. The movement on behalf of animals is a reform of society as a whole, which will ultimately lead to a new heaven and a new earth. Sperling especially emphasizes the millenarian aspects of the animal-protection movement. She follows historian Norman Cohn (1970), who outlines the characteristics of the millenarian ideal as collective, meaning enjoyed by a community of faithful; terrestrial, taking place on earth and not in heaven; imminent; total, resulting in an utter transformation of life on earth; and miraculous. If we omit the miraculous aspect, these characteristics describe the aspirations of many environmental and animal activists (Sperling 1988, 196). Sperling describes the "apocalyptic ecology" of British writer John Aspinall, who is often quoted by animal activists. Aspinall argues that we are at the brink of either ecological armageddon or a "partially restored" paradise: the choice is ours. He views animals as representatives of uncivilized nature, a state humans have given up to technology (Sperling 1988, 137–38). However, in a review of Sperling's book, Peter Singer (1989) cites this and similar comparisons as a way in which Sperling demonizes her subjects by branding them as fanatical.

A second set of metaphors in these books revolves around the dyads of intellectual/anti-intellectual and rational/irrational, whereby activists are viewed as irrational and emotional in opposition to rational science. Sperling is especially revealing in this regard. She concludes her book with an interview of Sandra Bressler, one of the founders of the California Biomedical Research Association, a proresearch group. Sperling's account emphasizes the "rational" quality of Bressler's demeanor and arguments, in contrast to what Bressler terms the "fanatical" and "irrational" animal activists. To Bressler, in Sperling's characterization, the issue is one of education: "[A]ccording to Sandra, it is crucial to apply reason to these emotionally charged issues. . . . Once the question is reframed, most people will not want to abolish animal research" (1988, 215).

Both sides in the debate agree that most people do not know what science is really about. Many sides, not just animal activists, have criticized the notion that science is privileged knowledge, not available to the ordinary onlooker. The recent debates on the "culture wars" stemming from Alan Sokal's article in *Social Text* (1996) illustrate, among other points, just how differently scientists and nonscientists perceive the practice of science. Demonizing is almost too easy. Some scientists believe their critics are simply ignorant; Jasper and Nelkin, for example, quote a scientist who refers to activists as "Yahoos" (1992, 129). On the other side, according to Sperling, activists view researchers as being in a constant state of denial, "on a kind of continuum of immorality" (1988, 154). Sperling goes on to describe the differing positions of scientists and activists as one of reason versus love (213). Activists are, by definition, anti-intellectual and fundamentally opposed to modernity, a description reminiscent of Fritz Ringer's, of the motivations for anti-Semitism in nineteenth-century Germany. Although Jasper and Nelkin are more nuanced, their account also emphasizes the emotional content of activists' appeals.

The idea of intellect versus instinct also relates to ideas about women. In certain versions of feminist philosophy, masculine/rational is opposed to feminine/intuitive. Women, it is argued, have different ways of understanding and are in addition more naturally compassionate and spiritual. In this view, nature itself is seen as female, and male science is a ravager and rapist. Jasper and Nelkin cite a pamphlet from a group called Feminists for Animal Rights, which claims that men find both women and animals to be irrational, instinctive, childish, and emotional, and that they are therefore treated in the same ways (1992, 53). That a majority of animal activists are women only enforces this conceptual gap. Sperling argues that the animal activists criticize women scientists most severely, viewing them as transgressing boundaries not only of moral behavior but of gender (1988, 13–14).

As the perceived originator of the gap between reason and emotion, between mind and body, Descartes remains demonized. Only the passionless Cartesian scientist could use terms such as "disposal" and "process" to discuss living creatures. Arrogant, even sadistic, scientists who privilege reason are themselves accused of acting without thought, "callous and devoid of compassion," in the words of an activist group.

The rhetorical gulf described in these two books will be difficult to breach. The language of demonization only makes this gulf wider; if each side views the other as being in a moral hinterland, little common ground for discourse can exist. I have tried here to give some historical insights into past and present rhetorical strategies. These strategies reveal how our underlying fears and misunderstandings have emerged in the discourse on animal protection. The various "rights" movements since the 1960s have demonstrated how powerful language can be and how everyday language and metaphors can indeed reveal our underlying assumptions. Can changes in language then change our views? In order to have meaningful conversations about the contentious issues surrounding animal research, we all need to be aware of the power of language both to demonize and to persuade.

Notes

I am grateful to Michael Osborne, Mark Schlenz, Harold Marcuse, Elise Robinson, and the members of the UCSB Research Focus Group on Animal-Human Relationships for their comments and assistance. I also wish to thank Sally Mitchell and Lori Williamson for sharing with me their research on Frances Power Cobbe, and members of the H-Albion list for answering queries.

1. See, for example, Augustine, book 9, chapter 5 (1982, 349).
2. Unless otherwise noted, translations of foreign-language works are by the author.
3. See also Guerrini 1989.
4. Rosenfield 1940 summarizes these views.
5. On *Naturphilosophie* see Coleman 1977; see also Lenoir 1982.
6. On the general topic of demonization, see Cohn 1975.

References

Arnauld, A., and P. Nicole. [1662] 1970. *La logique ou l'art de penser.* Reprint, Paris: Flammarion.
Augustine. 1982. *City of God.* Translated by Henry Bettenson. Harmondsworth, England: Penguin.
Bentham, J. 1789. *An introduction to the principles of morals and legislation.* London.
Bernard, C. [1878] 1974. *Lectures on the phenomena of life common to animals and plants.* Translated by H. E. Hoff, R. Guillemin, and L. Guillemin. Reprint, Springfield, Ill.: Thomas.

Birch, T., ed. 1772. Robert Hooke to Robert Boyle, 10 November 1664. In *The works of the honourable Robert Boyle,* 6:498. London.

Boyle, R. 1670. New pneumatical experiments about respiration. *Philosophical Transaction* 5:2044.

Brann, H. W. 1975. *Schopenhauer und das Judentum.* Bonn: Bouvier Verlag Hobert Grundmann.

Carmichael, J. 1992. *The satanizing of the Jews.* New York: Fromm.

Cobbe, Frances Power. 1891. *Jewish Chronicle,* 13 February.

Cohn, N. 1970. *The pursuit of the millennium.* Rev. ed. Oxford: Oxford University Press.

———. 1975. *Europe's inner demons.* New York: New American Library.

Coleman, W. 1977. *Biology in the nineteenth century.* Cambridge: Cambridge University Press.

Cottingham, J. 1978. "A brute to the brutes?" Descartes' treatment of animals. *Philosophy* 53:551–59.

Descartes, R. [1637, 1641] 1968. *Discourse on method and the meditations.* Translated by F. E. Sutcliffe. Reprint, Harmondsworth, England: Penguin.

———. [1649] 1971. Les passions de l'âme. (The Passions of the Soul.) Reprint, in *Oeuvres de Descartes.* Paris: J. Vrin.

Douglas, M. [1966], 1984. Reprint, *Purity and Danger.* London: Ark.

Fracassati, C. 1667. An account of some experiments of injecting liquors into the veins of animals. *Philosophical Transactions* 2:490.

French, R. D. 1976. *Anti-vivisection and medical science in Victorian society.* Princeton, N.J.: Princeton University Press.

Garber, D. 1993. Descartes and experiment in the *Discourse* and *Essays.* In *Essays on the philosophy and science of René Descartes,* edited by S. Voss. New York and Oxford: Oxford University Press.

Guarnieri, P. 1987. Moritz Schiff (1823–96): Experimental physiology and noble sentiment in Florence. In *Vivisection in historical perspective,* edited by N. A. Rupke, 105–24. New York: Croom Helm.

Guerrini, A. 1989. The ethics of animal experimentation in seventeenth-century England. *Journal of the History of Ideas* 50:391–407.

Hales, S. 1731. *Statical essays: Containing Haemastaticks.* London: W. Innys, T. Woodward, and J. Peele.

Haller, Albrecht von. [1753] 1935. On the sensible and irritable parts of the body. Translated 1755. Reprint, *Bulletin of the History of Medicine* 3:650–99.

Hull, D. S. 1969. *Film in the Third Reich.* Berkeley: University of California Press.

Jasper, J. M., and D. Nelkin. 1992. *The animal rights crusade: The growth of a moral protest.* New York: Free Press.

Katz, J. 1986. *The darker side of genius: Richard Wagner's anti-Semitism.* Hanover, N.H.: University Press of New England.

Lenoir, T. 1982. *The strategy of life.* Dordrecht: Reidel.

Linzey, A. 1987. *Christianity and the rights of animals.* New York: Crossroad.

MacIntyre, A. 1984. *After virtue*. Notre Dame, Ind.: Notre Dame University Press.

Manuel, D. 1987. Marshall Hall (1790–1857): Vivisection and the development of experimental physiology. In *Vivisection in historical perspective*, edited by N. A. Rupke, 78–104. New York: Croom Helm.

Nash, R. F. 1989. *The rights of nature*. Madison: University of Wisconsin Press.

Pernick, M. 1985. *A calculus of suffering: Pain, professionalism, and anaesthesia in nineteenth-century America*. New York: Columbia University Press.

Regan, T. 1983. *The case for animal rights*. Berkeley: University of California Press.

Rey, R. 1993. *Histoire de la douleur*. Paris: La Decouverte.

Ringer, F. 1969. *The decline of the German mandarins: The German academic community, 1890–1933*. Cambridge, Mass.: Harvard University Press.

Rosenfield, L. H. 1940. *From beast-machine to man-machine*. New York: Octagon.

Ruysch, F. [1665] 1964. Introduction to *Dilucidatio valvularum in vasis lymphaticis et lacteis*, edited by A. M. Luyendijk-Elshout. Amsterdam: Nieuwkoop.

Sax, B., and A. Arluke. 1995. The Nazi treatment of animals and people. In *Reinventing biology*, edited by R. Hubbard and L. Birke. Bloomington: Indiana University Press.

Sayers, D. L. 1923. *Whose body?* New York: Harper.

Schopenhauer, A. [1841] 1965. *On the basis of morality*. Translated by E. F. J. Payne. Indianapolis: Bobbs-Merrill.

Singer, P. 1975. *Animal liberation: A new ethics for our treatment of animals*. New York: New York Review.

———. 1989. Animal liberators: Research and morality. *New York Review of Books*, 2 February.

Smith, J. A., and K. M. Boyd, eds. 1991. *Lives in the balance*. Oxford: Oxford University Press.

Sokal, A. D. 1996. Transgressing the boundaries: Toward a transformative hermeneutics of quantum gravity. *Social Text* 46/47:217–52.

Sperling, S. 1988. *Animal liberators*. Berkeley: University of California Press.

Stern, F. 1961. *The politics of cultural despair*. Berkeley: University of California Press.

Trohler, U., and Andreas-Holger Maehle. 1987. Anti-vivisection in nineteenth-century Germany and Switzerland: Motives and methods. In *Vivisection in historical perspective*, edited by N. A. Rupke. London: Routledge.

Weiner, M. A. 1996. *Richard Wagner and the anti-Semitic imagination*. Lincoln: University of Nebraska Press.

Zambeccari, G. [1680] 1941. Experiments of Doctor Joseph Zambeccari concerning the excision of various organs from different living animals. Translated by S. Jarcho. *Bulletin of the History of Medicine* 9:311–31.

Cognitive Ethology, Deep Ethology, and the Great Ape/Animal Project

Expanding the Community of Equals

Marc Bekoff

Abstract: *In a previous chapter, Biller-Andorno demonstrated the need to understand the subjective lives of animals in order to understand our ethical obligations to them. In this chapter, Marc Bekoff, a noted biologist and bioethicist, concurs and describes how the field of classical ethology has been expanded in response to these important questions. He explains that modern ethology has expanded beyond the traditional areas of study, such as behavioral evolution, adaptation, and development, to include the private experience of animals. He then critiques a number of conceptualizations that have emerged as a consequence of this enlargement of focus. He explicitly criticizes notions of "higher" and "lower" species and the focus on primate mentality (as in the Great Ape Project) as an example of speciesism, a kind of animal racism. For example, he argues that trying to identify ape cognitive characteristics in canids, as in the question "Do dogs ape?" obscures the context of the animal's life. Bekoff reminds us that animals do what is necessary to live in their own worlds (or Umwelt, as Biller-Andorno would have it).*

By using examples of social play in dogs and antipredatory behavior in birds, Bekoff shows how, from a methodological point of view, a cognitive research program can be advanced. Readers get a glimpse of how observation in a field-research context can begin to implicate the cognitive existence of motives such as intentionality, belief, and negotiation. Bekoff also discusses the psychological perspective of the animal researcher, arguing that the researcher should adopt the perspective that animal study is a privilege, not a right. In opposition to the prevailing view, he further argues that a sense of bondedness between the human researcher and animal under study is the correct scientific and ethical stance. Bekoff reminds us that in research, animals under study often come to trust us and that this trust rightfully exerts a claim of respect and care.

Summary

In this essay I argue that the evolutionary and comparative study of non-human animal (hereafter, animal) cognition in a wide range of taxa can readily inform discussions about animal protection and animal rights. Although it is clear that there is a link between animal cognitive abilities and animal pain and suffering, I agree with Jeremy Bentham, who claimed in the eighteenth century that the real question does not concern whether individuals can think or reason but rather whether individuals can suffer. One of my major goals is to make the case that the time has come to expand the Great Ape Project (GAP) to the Great Ape/Animal Project (GA/AP) and to take seriously the moral status and rights of all animals by presupposing that all individuals should be admitted into a community of equals. I also argue that individuals count and that it is essential to avoid being speciesist cognitivists; it really doesn't matter whether "dogs ape" or whether "apes dog" when taking into account the worlds of different individual animals. We must resist narrow-minded primatocentrism and speciesism in our studies of animal cognition and animal protection and rights. Drawing lines into "lower" and "higher" species is a misleading speciesist practice that we should also vigorously resist; not only is line drawing bad biology but it also can have disastrous consequences for how animals are viewed and treated. Speciesist line drawing also ignores individual differences within species.

Expanding the Community of Equals and Doing Away with Speciesist Line Drawing: The Moral Importance of Individuals

> If the biological sciences have taught us one thing over the last one hundred years, it is that drawing all-or-nothing lines between species is completely futile. . . . [T]he cognitive and emotional lives of

animals differ only by degree, from the fishes to the birds to monkeys and to humans. (Fouts 1997, 372)

There is a profound, inescapable need for animals that is in all people everywhere, an urgent requirement for which no substitute exists.... Among the first inhabitants of the mind's eye, they are basic to the development of speech and thought. Later they play a key role in the passage to adulthood. Because of their participation in each stage of the growth of consciousness, they are indispensable to our becoming human in the fullest sense. (Shepard 1996, 3)

For the first time I understood how dumb a human can be in the presence of an intelligent animal. (Boone 1954, 21)

Happy animals make good science. (Poole 1997, 116)

In 1993 Cavalieri and Singer published *The Great Ape Project: Equality Beyond Humanity.* This important and seminal project has become widely known as the GAP. I was a proud contributor to the GAP (Bekoff 1993) and strongly supported its ambitious and major goal, namely, that of admitting Great Apes (including all humans) to the "community of equals," in which the following basic moral principles or rights, enforceable by law, are granted: (1) the right to life, (2) the protection of individual liberty, and (3) the prohibition of torture. When I wrote my brief essay I wanted to include all other nonhuman animals in this interdisciplinary appeal (in their epilogue Cavalieri and Singer agree with the importance of going beyond Great Apes), and now I believe that it is time to expand the GAP to the Great Ape/Animal Project, or the GA/AP, and to expand membership in the community of equals. In the GA/AP it will be presupposed that *all* individuals have the right to be included in the community of equals: nonprimates are important individuals in their own right, and neither they nor we need to engage in primate-envy in this "more-than-human world" (Abram 1996).

Although I realize that practical considerations made it important to start somewhere in the attempt to recognize the rights of nonhumans, I agree with Burghardt's concern about the original GAP. He states: "As one who believes the true test of our respect for other animals lies in our treatment of venomous snakes and large carnivores, I (too) am wary of a creeping speciesism inherent in the proposal set forth here" (Burghardt

1997b, 84). Of course, neither Burghardt nor I intend to undermine the intentions of the GAP; indeed, it is one of the most significant and ambitious worldwide pro-animal movements ever to arise with strong and ever-growing support from an interdisciplinary group of academics and nonacademics alike. However, given my own interests in animal cognition (cognitive ethology), I argue here that it is essential to avoid being *speciesist cognitivists* and that it really doesn't matter whether "dogs ape" or whether "apes dog" when taking into account the worlds of *individual* nonhuman animals.[1] For example, Marler concludes his review of social cognition in nonhuman primates and birds as follows: "I am driven to conclude, at least provisionally, that there are more similarities than differences between birds and primates. Each taxon has significant advantages that the other lacks" (1996, 22).

We must also avoid narrow-minded primatocentrism in our studies of animal cognition and animal protection and rights. Drawing lines into "lower" and "higher" species (see figure 1) is a misleading speciesist practice that we should vigorously resist. Line drawing is not only bad biology, but because it focuses on species differences, this practice can have disastrous consequences for how individual animals are viewed and treated. I and others have previously argued that it is individuals who are important.[2] For example, Rachels argues that "if it is wrong to use humans in experiments, then it is also wrong to use animals, *unless there are relevant differences between them that justify a difference in treatment*" (1990, 220, my emphasis). On this account careful attention must be paid to individual variations in behavior within species.

Animal Protection and
Researcher Responsibility: Shrimps and Chimps

The issues with which those interested in animal protection must deal are numerous, diverse, difficult, and extremely contentious, and interdisciplinary work is needed to try to come to terms with them (Bekoff 1998c, 2000b; see also <www.ethologicalethics.org>). Reasonable people with different backgrounds but common and deep interests in the protection of animals from human exploitation often disagree on the most basic issues and where different points of view can take them. Although some believe that we can do what we want to animals because there are no moral issues at stake (e.g.,

Speciesism	Nonspeciesism

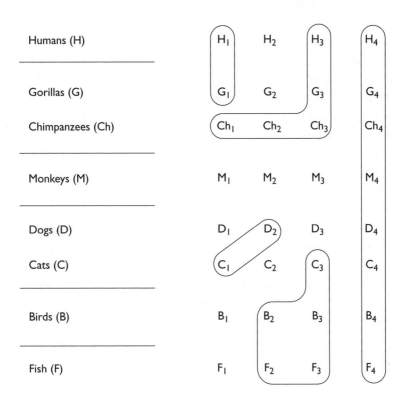

Humans (H)

Gorillas (G)

Chimpanzees (Ch)

Monkeys (M)

Dogs (D)

Cats (C)

Birds (B)

Fish (F)

Figure 1. A representation of speciesist and nonspeciesist perspectives.
(left) A speciesist representation of eight species (chosen for illustration). Lines dividing them into a linear hierarchy suggest, for example, that humans are "higher than" gorillas and chimpanzees and that monkeys are "higher than" dogs and cats. Speciesism provides a convenient way to make difficult decisions about which species may be used in different types of human activities, including research and teaching. However, this view does not pay serious attention to evolutionary continuity, and it deemphasizes individual variability. (right) A nonspeciesist representation of four individuals in each of the same eight species. Lines encircling various individuals ($H1$ and $G1$; $D2$ and $C1$) illustrate the point that individual characteristics "count." That is, it is possible that individual members of different species may be "equivalent" with respect to various traits or that individuals of a given species may possess characteristics that are exclusively theirs. Also, individuals of species that are typically thought to be "lower" than others may be more skilled in certain areas or experience pain, anxiety, and suffering more than individuals of species that are thought to be "higher." Nonspeciesist views argue against the use of species membership as the sole criterion for choosing which individuals should be used in various types of human activities.

White 1990),[3] these "moral privatists" (Jamieson 1985) fail to recognize that they are indeed taking a moral position. Fortunately, "moral privatists" are a rare breed.

It is essential to think seriously about what we do to individual animals when we study them "in the name of science"—how methods of study influence the lives of the animals and the nature of the data that are collected. I have studied various aspects of animal behavior and behavioral ecology (coyotes, wolves, and domestic dogs in captivity and in the field, Adélie penguins in Antarctica, and birds living in the front range of Colorado) for over 25 years, including animal cognition, and have also attempted to learn more about how the study of animal cognition can inform discussions of animal protection (Bekoff 1994; Bekoff and Jamieson 1991; see also Rollin 1992; Wemelsfelder 1997a,b). As a field researcher myself, I am interested in how field research can affect the lives of wild animals (Bekoff 1995a,b,c, 1997b; Bekoff and Jamieson 1996a). Our mere presence can influence the behavior of animals and disrupt such activities as feeding, mating, and care-giving behavior. Often we unknowingly interfere in the lives of the animals in whom we are interested. The guiding principles for all of my research are that (1) it is a privilege to study individual animals, (2) we should respect their dignity at all times, and (3) we must give primary attention to their views of the world. We must never forget that many animals give us so much in return. In many ways they make us human, facilitate contact with ourselves, teach us about trust and respect, and are there for us unconditionally. Respect and dignity are the least that we can give back to them.[4]

I have personally anguished many times over the different stances that I have adopted concerning animal use (Bekoff 1997a). In 1970 I dropped out of a graduate program because I did not want to continue to kill cats as part of my ongoing research. As I pondered my predicament, I recalled a bumper sticker carrying the message, "Join the Army; travel to exotic distant lands; meet exciting unusual people and kill them." I remember thinking that one could make a similar statement for some scientists: "Do science; engage in exciting research; meet wonderful animals and kill them." The very idea came to offend me, and I looked for other opportunities in which I could study the behavior of animals. Although I have not always lived up to my standards of conduct, I would never again engage in certain types of research, such as allowing coyotes to kill mice in staged encounters (Bekoff and Jamieson 1991). No one *has* to do research on animals, and no one else can assume responsibility for what each individual does.

In this essay I want to make some general comments about the study of animal minds by researchers who call themselves cognitive ethologists; I then briefly consider social play in various canids (members of the dog family) and antipredatory behavior in some birds to make the case that many animals have very active minds—they make plans, have desires, and have beliefs. I also want to stress that we must not be speciesist cognitivists —we should strive to study individual animals to learn more about their lives in their own worlds and their abilities to feel pain and to suffer psychologically and physically. In the absence of suitable criteria and empirical data for making comparative claims about smartness or intelligence, we should be careful about making such statements as "apes are smarter than monkeys or dogs" and drawing moral conclusions, for each can do things the other cannot. Along these lines Tomasello and Call conclude in their comprehensive review of primate cognition that "[t]he experimental foundation for claims that apes are more intelligent than monkeys is not a solid one, and there are few if any naturalistic observations that would substantiate such broad-based, species-general claims" (1997, 399–400). *Smart* and *intelligent* are loaded words and often are misused: dogs do what they need to do to be dogs—they are dog-smart in their own ways; and monkeys do what they need to do to be monkeys—they are monkey-smart in their own ways; and neither is necessarily smarter than the other. The misunderstanding and misapplication of the notions of smartness and intelligence can have significant and disastrous consequences for animals.

This line of reasoning does not mean that we cannot talk about whether chimps are smarter than shrimps, for example. Of course we can, especially when practical matters arise. However, we must be very clear about why we might feel comfortable with the claim that chimps on average are smarter than shrimps. We need to be clear about the criteria we use to make comparative statements about smartness and intelligence—what we mean when and if we claim that chimps' social lives are more complex or that chimps are able to solve more complex or difficult problems than are shrimps. The main point is that whereas most people would probably not have trouble deciding to harm or to kill a shrimp rather than a chimp if forced to make a choice, their decision should not be made summarily or casually. Drawing moral boundaries at the species level using some set of average species-typical characteristics is fraught with difficulties. As I have already mentioned, individual differences within species need to be given careful consideration in moral deliberations; and even within the confines of moral individualism, decisions about how individuals may be used and abused are extremely difficult.

Cognitive Ethology

Currently, there is growing interest in cognitive ethology, or the study of animal minds, by scientists, philosophers (Allen and Bekoff 1997; Bekoff 1998a; Bekoff and Jamieson 1996b; Tomasello and Call 1997), and lay people alike. Animal minds and sentience are being taken seriously by many people (and should be taken seriously by all people) not only to learn more about the other living beings with whom we share this planet but also because it is clear that a link exists between cognitive abilities and the ability to experience pain and suffer. However, we must remember, as the philosopher Jeremy Bentham claimed in the eighteenth century, that the real question does not deal with whether individuals can think or reason but rather with whether individuals can suffer: "But a full-grown horse or dog is beyond comparison a more rational, as well as a more conversable animal, than an infant of a day, or a week, or even a month, old. But supposed they were otherwise, what would it avail? The question is not Can they *reason?* nor Can they *talk?* but Can they *suffer?*" ([1789] 1948).

My thesis therefore is that when we are unsure about an individual's ability to reason or to think, we should assume that they can in their own ways; and certainly when we are uncertain about an individual's ability to experience pain and to suffer, we must assume that they can. We must err on the side of the animals.

Because I write here about various canids and birds, and do not emphasize work done on primates, I will start with a few quotations that set the stage for my appeals to broaden our taxonomic horizons—to go beyond the great apes and other nonhuman primates and to conduct broad comparative and evolutionary studies of animals in the field. The psychologist David Premack has claimed that cognition is for the most part a laboratory phenomenon at least in chimpanzees: "[C]ognition is not exclusively a field phenomenon; it can take place in the laboratory. Indeed, in the case of chimpanzees, advanced cognition would appear to be largely a laboratory phenomenon. For only the chimpanzee who has been specially trained and exposed to the culture of a species more evolved than it-self—shows analogical reasoning. . . . [A]dvanced cognition, such as ana-logical reasoning, is confined to the laboratory" (Premack 1988, 171–72).

I and many others think Premack is wrong (Allen and Bekoff 1995, 1997; Bekoff and Allen 1997; Byrne 1995; de Waal and Aureli 1996), and a large database exists that supports the view that many animals perform be-havior patterns in the field that would uphold even a minimalist attribu-tion of cognitive abilities. Indeed, Sue Savage-Rumbaugh and her col-leagues (1996) are pursuing this question in their work on trail following by bonobos. There also is a tendency to dismiss prematurely the cognitive skills of nonprimates (see Beck 1982; Marler 1996). For example, Mason claims, "On the basis of findings such as those reviewed in this paper, I am persuaded that the apes and man have entered a cognitive domain that sets them apart from all other primates" (1979, 292–93). Along these same lines are some of the more recent claims that Richard Byrne makes in his oth-erwise excellent book *The Thinking Ape* (1995, my emphases added):

> It *seems* that the great apes, especially the common chimpanzee, can attribute mental states to other individuals; but no other group of ani-mals can do so—apart from ourselves, and perhaps cetaceans. (146)

> This contrasts with the findings on understanding of beliefs, attri-bution of intentions, and how things work—where a sharp discon-tinuity is *implied* between great apes and all other animals. (154)

> Of course, until similar painstaking work is done with monkeys, we *cannot* argue that only apes have such abilities and no-one has yet risked the huge expenditure of time and money to find out. (172)

> And with independent evidence that gorillas can attribute intentions, the apparently anthropomorphic account becomes the more parsi-monious one. (161)

I will return to some of these points later, but it is clear that Byrne qualifies his broad generalizations and admits that the broad comparative research that needs to be done has not yet been conducted. Furthermore, even much of the work that has been done on nonhuman primates has involved very few individuals representing only a few of the many extant species.

What about the minds of animals other than humans? Is there any-thing in them? If there is, how can we ever know what it is? If they do not experience pain and suffering in the ways that humans do, how can we ever be sure that they really do experience pain and suffering? While most people have no doubt that there is a lot in the minds of many ani-mals, they also recognize that we will only learn about what is there with careful studies. Such studies will be difficult, but difficult does not mean impossible—we just need some patience. In addition, these studies will have to be interdisciplinary, involving at least biologists, psychologists, an-thropologists, and philosophers. The results of these endeavors will allow us to learn more about animal cognition and consciousness and possible connections among cognition, consciousness, and an individual's ability to feel pain and suffer.

However, not everyone agrees that even some animals can have con-scious experiences or that their pains are morally relevant. For example, Peter Carruthers claims that the experiences of all animals are all non-conscious. He writes: "I shall assume that no one would seriously main-tain that dogs, cats, sheep, cattle, pigs, or chickens consciously think things to themselves. The experiences of all these creatures [are of] the non-conscious variety" (1989, 265). Carruthers then claims:

> Similarly then in the case of brutes: since their experiences, includ-ing their pains, are nonconscious ones, their pains are of no imme-diate moral concern. Indeed since all the mental states of brutes are nonconscious, their injuries are lacking even in indirect moral con-cern. Since the disappointments caused to the dog through posses-sion of a broken leg are themselves nonconscious in their turn, they, too, are not appropriate objects of our sympathy. Hence, neither the pain of the broken leg itself, nor its further effects upon the life of the dog, have any rational claim upon our sympathy. (268)

And finally, he concludes, "And it also follows that there is no moral crit-icism to be leveled at the majority of people who are indifferent to the pains of factory-farmed animals, which they know to exist but do not themselves observe" (269).

However, as Lynch (1994) has noted, the mere possibility that we can account for some pain as a type of unconscious perception does not suffice to establish that animal pain is most plausibly interpreted in this way. Others also have mounted serious objections to Carruthers's views for a wide variety of reasons (Boonin-Vail 1994; Jamieson and Bekoff 1992; E. Johnson 1991; Pluhar 1993a,b, 1995; Robinson 1997), and Carruthers has not been very successful in his attempts to dismiss categorically the possibilities of conscious experiences and morally relevant pain in non-humans. (Carruthers is softening his moral position because he does not want to push the entailment between the having of nonconscious experiences and ethical implication [personal communication with the author, 4 July 1997].) Yet Carruthers is not alone. Consider Bermond's less (but still overly) restrictive view that "the claims for suffering in animal species, other than in anthropoid apes and possible [*sic*] dolphins, are incorrectly substantiated. Such claims are the products of anthropomorphic projections" (1997, 138).

The Evolutionary, Comparative, and Ecological Study of Animal Minds

The interdisciplinary science of cognitive ethology mainly involves claims about the evolution of cognitive processes.[5] Because behavioral abilities have evolved in response to natural selection pressures, ethologists favor observations and experiments on animals in conditions that are as close as possible to the natural environment in which selection occurred. In addition to situating the study of animal cognition in a comparative and evolutionary framework, cognitive ethologists also maintain that field studies of animals that include careful observation and experimentation can inform studies of animal cognition and that cognitive ethology need not be brought into the laboratory to make it respectable. Furthermore, because cognitive ethology is a comparative science, cognitive ethological studies emphasize broad taxonomic comparisons and do not focus on a few select representatives of limited taxa. Cognitive psychologists, in contrast to cognitive ethologists, typically work on related topics in laboratory settings and do not emphasize comparative or evolutionary aspects of animal cognition. When cognitive psychologists do make cross-species comparisons, they are generally interested in explaining different behavior patterns in terms of common underlying mechanisms; ethologists, in common with other biologists, are often more concerned with the diversity of solutions that living organisms have found for common problems.

Comparative cognitive ethology is an important extension of classical ethology because it explicitly licenses hypotheses about the internal states of animals in the tradition of classical ethologists such as Nobel laureates Niko Tinbergen and Konrad Lorenz. However, although ethologists such as Lorenz and Tinbergen used terms such as *intention movements,* they used them quite differently from how they are used in the philosophical literature. "Intention movements" refers to preparatory movements that may communicate what action individuals are likely to do next; they do not necessarily refer to their beliefs and desires, although one could suppose that some individuals did indeed want to fly and believed that if they moved their wings in a certain way they would fly. This distinction is important because the use of such terms does not necessarily add a cognitive dimension to classical ethological notions, although it could. Indeed, Tinbergen once wrote: "Concepts such as 'play' and 'learning' have not yet been purged completely from their subjectivist, anthropomorphic undertones. Both terms have not yet been satisfactorily defined objectively, and this might well prove impossible" (1963, 13).

In his early work Tinbergen (1989; 1963) identified four overlapping areas with which ethological investigations should be concerned, namely, evolution (phylogeny), adaptation (function), causation, and development (ontogeny). His framework also is useful for those interested in animal cognition (Allen and Bekoff 1997; Jamieson and Bekoff 1993). Burghardt has suggested adding a fifth area, private experience. According to Burghardt, "The fifth aim is nothing less than a deliberate attempt to understand the private experience, including the perceptual world and mental states, of other organisms. The term private experience is advanced as a preferred label that is most inclusive of the full range of phenomena that have been identified without prejudging any particular theoretical or methodological approach" (1997a, 276).

Burghardt also notes that using the term *cognitive ethology* for the fifth area is not radical enough and that there may be some historical baggage associated with the use of this term. For instance, it is frequently associated with only the study of animal consciousness. However, cognitive ethology is much more than the study of animal consciousness (Allen and

Bekoff 1997; Bekoff 1998a, 2000a), although many do not mark this distinction (see Bekoff and Allen 1997) and summarily dismiss the field.

The modern era of cognitive ethology and its concentration on the evolution and evolutionary continuity of animal cognition is usually thought to have begun with the appearance of Donald R. Griffin's book *The Question of Animal Awareness: Evolutionary Continuity of Mental Experience* (1981), first published in 1976. Griffin's major concern was to learn more about animal consciousness and its evolutionary/ancestral variants, and then, as now, Griffin wanted to come to terms with the difficult question, "What is it like to be a particular animal?" Although Griffin was mainly concerned with the phenomenology of animal consciousness, it is only one of many important and interesting aspects of animal cognition. Indeed, because of its broad agenda and wide-ranging goals, many view cognitive ethology as being a genuine contributor to cognitive science in general. For those who are anthropocentrically minded, it should be noted that studies of animal cognition can also inform, for example, inquiries into human autism (Baron-Cohen 1995).

Methods of Study: Naturalizing and Individualizing the Study of Cognition

The methods for answering questions in the five areas that Tinbergen and then Burghardt identified for ethological investigations—evolution, adaptation, causation, development, and private experience—vary, but all begin with careful observation and description of the behavior patterns that are performed by the animals under study. The information afforded by these initial observations allows a researcher to exploit the animal's normal behavioral repertoire to answer questions about the evolution, function, causation, and development of the behavior patterns that are performed in various contexts, and perhaps also private experience. Naturalizing and individualizing the study of animal cognition and animal minds in the laboratory and in the field should lead to a greater appreciation for the cognitive skills of animals living under natural conditions.

Basically, there are no large differences between methods used to study animal cognition and those used to study other aspects of animal behavior.[6] Differences lie not so much in what is done and how it is done but in how data are explained. Likewise, the main distinction between cognitive ethology and classical ethology lies not in the types of data collected but in the understanding of the conceptual resources that are appropriate for explaining those data (Allen and Bekoff 1997).

In studies of behavior it is important to know as much as possible about the sensory world of the animals being studied. Experiments should not be designed that ask animals to do things that they cannot do because they are insensitive to the experimental stimuli, unmotivated by the stimuli, or unable to perform the required action. The relationships between normal ecological conditions and differences between the capabilities of animals to acquire, process, and respond to information is the domain of a growing field called "sensory ecology." Ethologists frequently ask, "What is it like to be the animal under study?" and develop a keen awareness of the senses that the animals use singly or in combination with one another. It is highly unlikely that individuals of any other species sense the world the same way we do, and it is unlikely that even members of the same species sense the world identically all of the time, and it is important to remain alert to the possibility of individual variation.

Stimulus Control and Impoverished Environments

Although carefully conducted experiments in the laboratory and in the field often can control for the influence of variables that may affect the expression of behavioral responses, there is usually a possibility that the influence of some variable(s) cannot be accounted for. Field studies may be more prone to a lack of control because the conditions in which they are conducted are inherently more complex and less controllable, and this presents one of the greatest challenges for the study of cognition (de Waal and Aureli 1996, 83 ff.).

The excellent cognitive ethological field research of Cheney and Seyfarth (1990) on the behavior (e.g., communication and deception) and minds of vervet monkeys illustrates such a concern for control in investigations. In their studies of the attribution of knowledge by vervets to each other, Cheney and Seyfarth played back vocalizations of familiar individuals to other group members. The researchers were concerned about their inability to eliminate "all visual or auditory evidence of the [familiar] animal's physical presence" (230). Actually, this inability may not be problematic if the goal is to understand "how monkeys see the world." Typically, in most social situations the physical presence of individuals and access to stimuli from different modalities may be important to consider. Vervets, other nonhumans, and humans may attribute mental states using a combination of variables that are difficult to separate experimentally. Negative or inconclusive experimental results concerning vervets' or other animals' attribution of mind to other individuals may stem from impoverishing their normal environment, by removing information that they normally use in attribution. Researchers might also be looking for complex mechanisms involved in the attribution of minds to others and could overlook relatively simple means for doing so. Just because an animal does not do something does not mean that it cannot do it (assuming that what we are asking the animal to do is reasonable, that is, within their sensory and motor capacities). Thus, insistence on absolute experimental control that involves placing and maintaining individuals in captivity and getting them accustomed to test situations that may be unnatural may greatly influence results. And these sorts of claims, if incorrect, can wreak havoc on discussions of the evolutionary continuity of animal cognitive skills. Cheney and Seyfarth recognize some of these problems in their discussion of the difficulties of distinguishing between alternative explanations maintaining either that a monkey recognizes another's knowledge or that a monkey monitors another's behavior and adjusts his or her own behavior to the other.

Although control may be more of a problem in field research than in laboratory work, cognitive ethologists should certainly not abandon field work. Cognitive ethologists and comparative or cognitive psychologists can learn important lessons from one another. On the one hand cognitive psychologists who specialize in highly controlled experimental procedures can teach something about the importance of control to those cognitive ethologists who do not perform such research. On the other hand those who study humans and other animals under highly controlled and often contrived and impoverished laboratory conditions can broaden their horizons and learn more about the importance of more naturalistic methods; they can be challenged to develop procedures that take into account possible interactions among stimuli within and between modalities in more naturalistic settings. The use of single tests relying primarily on one modality—for example, vision—for comparative studies represents too narrow an approach. Ultimately, all types of studies should be used to exploit the behavioral flexibility or versatility of the animals under study.

Two Case Studies:
Social Play and Antipredatory Behavior

Social Play and Negotiation and Cooperation

> To return to our immediate subject: the lower animals, like man, manifestly feel pleasure and pain, happiness and misery. Happiness is never better exhibited than by young animals, such as puppies, kittens, lambs, etc., when playing together, like our own children. Even insects play together, as has been described by that excellent observer, P. Huber, who saw ants chasing and pretending to bite each other, like so many puppies. (Charles Darwin [1871] 1936, 448)

In addition to social play and antipredator behavior, one could also discuss topics such as language acquisition and use, tool use, mirror image recognition, and deception; many examples can be found in Allen and Bekoff (1995, 1997); Bekoff and Jamieson (1996b); Cheney and Seyfarth (1990); Griffin (1992); Ristau (1991b), and references therein. Although Griffin has included the results of many excellent studies of the possibility of language in nonhuman primates, cetaceans, and birds in his broad discussions of animal minds, they do not squarely fall within the primary domain of cognitive ethology as I envision it: the study of natural behaviors in natural settings from an evolutionary and ecological perspective. (Of course, this is not to discount the importance to cognitive ethology of research on captive animals.) Only future research will tell if the behavior of the few captive individuals who have been intensively studied in "language studies" (and those captive individuals observed in other endeavors) is related to the behavior of wild members of the same species (e.g., Savage-Rumbaugh et al. 1996) or if the data from captive animals are more an important demonstration of behavioral plasticity and behavioral potential.

Social play is an easily recognizable behavior with which most people are familiar. When animals play they look like they're having fun, they run around with a loose gait, they repeatedly try to get others to play with them, and when they can't get another animal to play, they try to get others to play with them or they play by themselves—they chase their tails or chase a ball or play with a food bowl. Play looks like fun—and it certainly is. Animals typically play when they are nonstressed, and many have concluded that something is profoundly wrong when dogs and other animals stop playing.

The study of animal play not only provides access into animals' minds but can also provide information that is important in considerations of the treatment to which animals are subjected. Consider the following description of two dogs, Jethro and Rosie.

Jethro runs toward Rosie, stops immediately in front of her, crouches on his forelimbs (bows), wags his tail, barks, and immediately lunges at her, bites her scruff and shakes his head rapidly from side to side, works his way around to her backside and mounts her, jumps off, does a rapid bow, lunges at her side and slams her with his hips, leaps up and bites her neck, and runs away. Rosie takes wild pursuit of Jethro and leaps on his back and bites his muzzle and then his scruff, and shakes her head rapidly from side to side. They then wrestle with one another and part, only for a few minutes. Jethro walks slowly over to Rosie, extends his paw toward her head, and nips at her ears. Rosie gets up and jumps on Jethro's back and bites him and grasps him around his waist. They then fall to the ground and wrestle with their mouths.

This description of a play encounter between two dogs (which could have as well occurred between other canids, felids, nonhuman primates, or humans) shows that when they engage in social play they perform behavior patterns that are used in other contexts, including aggression, reproduction, and predation. They and other animals, including humans, also use actions that are important for initiating and maintaining play, in this case "bows" (dogs crouch on their forelimbs and may wag their tail and bark). Social play in animals is usually a cooperative turn-taking venture, which leads to an important question: "How do animals negotiate cooperative agreements?" For example, how might dogs know that their playmates want to play with them, not fight with them? And how can one dog tell another, "I want to play with you" or "I want to play with you no matter what I just did or am about to do"? To answer these questions I studied where play bows are placed in ongoing play—how various canids use bows during social play.

I found that bows are performed when the signaler wants to communicate a specific message about his or her desires or beliefs (Bekoff 1995c). Dogs, wolves, and coyotes did not perform bows randomly; rather, they used bows more often immediately before or immediately after biting accompanied by head shaking than at other places in a play bout—statistically significant trends. Biting accompanied by head shaking could (easily) be misinterpreted, and the animals seemed to want to reduce the likelihood that this would happen.

How important is it to negotiate play and to agree that play is the name of the game? Very much so. When animals play, they for the most part borrow behavior patterns from other contexts, and individuals need to be able to tell one another that they do not want to eat, fight with, or mate with the other individual(s), but rather that they want to play with them. In most species in which play has been observed, specific actions have evolved that are used to initiate play. These actions seem to function in negotiations between participants, the result of which is that they come to an agreement to engage in cooperative play rather than aggression or predation, for example. There is no solid evidence that animals invite others to play and then exploit them. Furthermore, researchers have also observed self-handicapping (for example, controlling the intensity of bites) and role reversals (dominant individuals assuming submissive roles only in play) in many species, including nonprimates. These behavior patterns are often used to support claims that at least some nonhuman primates are conscious or have a concept of self.

Social-play behavior is but one activity in which animals seem to have desires and beliefs and to make plans. Dogs show their desire to play by performing a bow, believing that if they do it then play will occur because the other animal will agree to play. This involves making plans for the future—I want to play with you. In addition, dogs may plan to bite their partners vigorously in a moment but tell their partner that they do not mean what they are going to do or did not mean what they just did. Although we cannot be sure that two dogs, for example, have beliefs about the effects of their behavior on another individual(s), some data do suggest this possibility. For example, suppose we wanted to know why Rosie permitted Jethro to nip at her ears. One explanation may be that Rosie believes Jethro is playing. And perhaps Jethro believes that Rosie believes that Jethro is playing. Providing answers to questions such as these is one of the challenges of research in animal cognition.

To sum up, at least some canids (and most likely other mammals) cooperate when they engage in social play, and they may negotiate and agree to engage in these ongoing cooperative ventures by sharing their intentions. In general, animals engaged in social play use specific signals to modulate the effects of behavior patterns that are typically performed in other contexts but whose meaning is changed in the context of play. These signals often relate flexibly to the occurrence of events in a play sequence that might violate expectations within that sequence.

All in all, it is highly likely that a detailed consideration of social play will help promote the development of more sophisticated theories of intentionality, representation, communication, and consciousness, from which we will learn more about individual beliefs, desires, abilities to make plans, and expectations about the future. In turn, all of these capacities connect closely to how animals suffer at the hands of humans—how they perceive and feel about the situations in which they currently find themselves or will find themselves, and how they react to them.

Antipredator Behavior in Birds

Ristau (1991a) studied injury feigning in piping plovers (the broken-wing display) and wanted to know if she could learn more about deceptive injury feigning if she viewed the broken-wing display as an intentional or purposeful behavior ("the plover wants to lead the intruder away from her nest or young") rather than a hard-wired reflexive response to the presence of a particular stimulus, a potentially intruding predator. Ristau studied the direction in which birds moved during the broken-wing display, how they monitored the location of the predator, and the flexibility of the response. She found that birds usually performed the display in a direction that would lead an intruder who was following them further away from the threatened nest or young; she also discovered that birds monitored the intruder's approach and modified their behavior in responses to variations in the intruder's movements. These and other data led Ristau to conclude that the plover's broken-wing display lent itself to an intentional explanation—that plovers purposely lead intruders away from their nests or young and modify their behavior in order to do so.

In another study of antipredator behavior in birds, I found that western evening grosbeaks modified their vigilance or scanning behavior depending on the way in which individuals were positioned with respect to one another (Bekoff 1995d, 1996). Grosbeaks and other birds often trade off scanning for potential predators and feeding; essentially (and oversimplified), some birds scan while others feed, and some birds feed when others scan. Thus, I hypothesized that individuals want to know what others are doing and learn about others' behavior by trying to watch them. My study of grosbeaks showed that when a flock contained four or more birds, large changes in scanning and other patterns of behavior occurred that seemed to be related to ways in which grosbeaks attempted to gather information about other flock members. Birds arranged in a circular array so that they could see one another easily compared with birds arranged in a line, which made visual monitoring of flock members more difficult, showed different behaviors. Birds who had difficulty seeing one another were more vigilant, changed their head and body positions more often, reacted to changes in group size more slowly, showed less coordination in head movements, and showed more variability in all measures. The differences in behavior between birds organized in circular arrays and birds organized in linear arrays could be best explained by accounting for individuals' attempts to learn, via visual monitoring, about what other flock members were doing. This may say something about if and how birds attempt to represent their flock, or at least certain other individuals, to themselves. It may be that individuals form beliefs about what others are most likely doing and predicate their own behavior on these beliefs. I have argued that cognitive explanations are simpler and less cumbersome than noncognitive rule-of-thumb explanations (e.g., "scan this way if there are this number of birds in this geometric array" or "scan that way if there are that number of birds in that geometric array" (Bekoff 1996). Noncognitive rule-of-thumb explanations did not seem to account for the flexibility in animals' behavior as well or as simply as explanations that appealed to cognitive capacities of the animals under study.

The Future

Methodological pluralism is essential: species-fair methods need to be tailored to the questions and animals under consideration, and one must always consider competing hypotheses and explanations. We are a long way from having an adequate database from which stipulative claims about the taxonomic distribution of various cognitive skills or of theories of mind can be put forth. Therefore, questions such as "Do mice ape?" or "Do apes mice?" are not very useful for this and for other reasons.

Those interested in animal cognition should resist temptations to be speciesist cognitivists who make sweeping claims about the cognitive abilities (or lack thereof) of all members of a given species. A concentration on individuals and not on species should form an important part of the agenda for future research in cognitive ethology. There is much individual variation in behavior within species, and generalizations about what an individual ought to do because it is classified as a member of a given species must be taken with great caution. Furthermore, people often fail to recognize that in many instances sweeping generalizations about the cognitive skills (or lack thereof) of species and not of individuals are based on small data sets from a limited number of individuals representing few taxa, individuals who may have been exposed to a narrow array of behavioral challenges. The importance of studying animals under field conditions cannot be emphasized too strongly. Field research that includes careful and well-thought-out observation, description, and experimentation that does not result in mistreatment of the animals is extremely difficult to duplicate in captivity. Although it may be easier to study animals in captivity, such animals must be provided with the social and other stimuli to which they are exposed in the field. In some cases this might not be possible.

What people believe about the cognitive capacities of nonhumans informs how they think about animal welfare—different views dispose a person to look at animals in particular ways. Ascribing intentionality and other cognitive abilities to animals is not moot if there are moral consequences, and there are. Nonetheless, individuals' abilities to experience pain, suffer, or experience anxiety that may threaten their well-being and violate their rights provide more compelling reasons to grant them moral status and to treat them with respect than does their ability to perform actions that submit to cognitive explanations (i.e., that they have memories of past events, are aware of their surroundings, have the ability to think about things that are absent, or can have beliefs or desires and be able to make future plans). This point needs to be stressed because, at least at the moment, it seems impossible to come up with any rigorous criteria that lead to the conclusion that specific and perhaps species-typical cognitive abilities are morally relevant, whereas others are not. Line drawing thus becomes at best a reckless activity with potentially serious consequences for the animals who find themselves "below" some arbitrary cut-off point for moral consideration.

So "do dogs ape?" or "do fish dog?" or "do dogs fish?" or "do apes dog?" Here are some contrasting views:

There is a growing consensus that because of homology in behaviour and nervous structure all vertebrates, thus also fish, have subjective experiences and so are able to suffer. (Verheijen and Flight 1997, 362)

What is imperative for chimpanzees, highly developed mammals biologically akin to ourselves, is inappropriate for cattle or lower species. (Leahy 1996, 190)

Still, there is a huge difference between our minds and the minds of other species, a gulf wide enough even to make a moral difference. (Dennett 1995, 371)

And one of the most arrogant of these ideas was the conceit that while I, because of my "divinely bestowed superiority," was fully qualified to communicate important thought down to the animals, the animals, because of their "divinely bestowed inferiority," were able to communicate little of real value up to me. (Boone 1954, 73)

People often ask whether "lower" nonhuman animals such as fish or dogs perform sophisticated patterns of behavior that are usually associated with "higher" nonhuman primates—"Do fish ape?" or "Do dogs ape?" In my view these are misguided questions, as is the question "Do apes dog?" because animals have to be able to do what they need to do in order to live in their own worlds. This type of speciesist cognitivism also can be bad news for many animals. If an answer to this question means that there are consequences in terms of the sorts of treatment to which an individual is subjected, then we really need to analyze the question in great detail.[7] It is important to accept that while there are species differences in behavior, behavioral differences in and of themselves may mean little for arguments about the rights of animals. As I have noted, the use of the words *higher* and *lower* and activities such as line drawing to place different groups of animals above and below others are extremely misleading and fail to take into account the lives and the worlds of the animals themselves. Their lives and worlds are becoming increasingly accessible as the field of cognitive ethology matures. Irresponsible use of these words can also be harmful for many animals. It is disappointing that a recent essay on animal use in a widely read magazine, *Scientific American,* perpetuates this myth—this ladder view of evolution—by referring to animals "lower on the phylogenetic tree" (Mukerjee 1997, 86). In the same issue of *Scientific American* we are told, "In my opinion, the arguments for banning experiments on animals—that there are empirically and morally superior alternatives—are unpersuasive" (Rennie 1997, 4). Poole also writes about "higher animals, like us" (1997, 117).[8] There are a number of objections to hierarchical ladder views of evolution, including that a single "ladder view" of evolution does not take into account animals with uncommon ancestries (Crisp 1990; see also Pinker 1994) and that there are serious problems deciding which criteria for moral relevance should be used and how evaluations of these criteria are to be made, even if one was able to argue convincingly for the use of a single scale (Bekoff 1992). To be sure, ladder views are speciesistic.

As I previously noted, some primatologists write as if only some non-human primates along with human primates have theories of mind. To dismiss the possibility that at least some nonprimates are capable of having a theory of mind requires the collection of much more data and consideration of existing data about intentionality in nonprimates. Furthermore, primatocentric claims are based on very few comparative data derived from tests on a small numbers of nonhuman primates who may not be entirely representative of their species. The range of tests that have been used to obtain evidence of intentional attributions is also extremely small, and such tests are often biased toward activities that may favor apes over monkeys or members of other nonprimate species. However, there is evidence that mice can outperform apes on some imitation tasks (Whiten and Ham 1992). These data do not make mice "special," and I am sure few would claim that these data should be used to spare mice and exploit monkeys. Rather, these results show the importance of investigating the abilities of various organisms with respect to their normal living conditions. Accepting that there are species differences in behavior, and that behavioral differences in and of themselves may mean little to arguments about the rights of animals, is important, for speciesist cognitivism can be bad news for many animals.

Deep Ethology and Animal Protection: Expanding the GAP to the GA/AP

> Surely it is our animal nature that recognizes the divinity of the natural world in all its mystery and beauty, despite the distressing habits and limited perception that afflict our species. So perhaps our hope of redemption lies in the fact that we are animals, not that we are people. (Thomas 1996, 126)

> I have often wondered what science might look like if, instead of having animals in numbered lots, they were treated respectfully as individuals—what science might have become had its history been different, had it not relied on distancing ourselves from nature. (Birke 1997, 55)

> Science no longer occupies the privileged and unassailable position that it once did. People increasingly question the benefits of "progress" in extending life, engineering the human and animal genome, and developing new reproductive and biomedical technologies. Science, moreover, is largely a publicly funded activity. Appropriately, accountability is the new watchword, and public education and consensus-building are the new goals. (Mench 1996, 9)

> Without the exchange of a sound or a gesture between us, each had perfectly understood the other. I had at last made contact with that seemingly lost universal silent language which, as those illuminated ancients pointed out long ago, all life is innately equipped to speak with all life whenever minds and hearts are properly attuned. (Boone 1954, 72)

From a humane point of view, there is no question that the lucky animals are those that are killed by people, whether it be by humane slaughter, a hunter, a car accident, or euthanasia by a humane organization or researcher. (Howard 1994, 202)

Yet an environmental ethic that excludes the bulk of life [the lower (*sic*) vertebrates, invertebrates, and plants], as well as the systems which render all this existence possible, represents too narrow a basis for valuing and protecting nature's diversity. (Kellert 1997, 205)

It is important to talk to the animals and let them talk to us; these reciprocal conversations should allow us to see the animals for who they are. Boone realized that as he was coming to learn more and more about his canid companion Strongheart, "I had never actually *seen* a dog! I had merely *looked* at dogs, without being able really to *see* one of them" (1954, 60). Likewise, Gluck (1997), in stressing the importance of considering what we do to animals from the perspective of the animal, emphasizes the need to go beyond science and to see animals as who they are. Our respect for animals must be motivated by who they are and not by who we want them to be in our anthropocentric scheme of things. As Taylor notes, a switch away from anthropocentrism to biocentrism, in which human superiority comes under critical scrutiny, "may require a profound moral reorientation" (1986, 313). So be it.

We are still a long way from having an adequate database from which stipulative claims about the taxonomic distribution of various cognitive skills, or about the having of a theory of mind, can be put forth with any degree of certainty. Furthermore, we still have little idea about the phylogenetic distribution of pain and suffering in animals. We can only hope that adequate funding will be available so that these important studies can be pursued rigorously.

With respect to possible links between the study of animal cognition and the protection of innocent nonconsenting animals, I believe that a deep reflective ethology is needed to make people more aware of what they do to nonhumans and to make them aware of their moral and ethical obligations to animals. We must enter into intimate and reciprocal relationships with all beings in this more-than-human world (Abram 1996). In many circles it simply is too easy to abuse animals. I use the term *deep reflective ethology* to convey some of the same general ideas that underlie the deep ecology movement, which asks that people recognize that they not only are an integral part of nature but also have unique responsibilities to nature. Most people who think deeply about the troubling issues surrounding animal welfare would agree that the use of animals in research and education, for amusement, and for food needs to be severely restricted and in some cases simply stopped. Those who appeal to the "brutality of nature" to justify some humans' brutal treatment of nonhumans fail to see that animals are not moral agents and cannot be held responsible for their actions as being "right" or "wrong" or "good" or "bad" (Bekoff and Hettinger 1994). If animals were to be viewed as moral agents (rather than as moral patients), there are a number of cognitive abilities that are correlated with the ability to make moral judgments, the possession of which would make animal abuse even more objectionable. It is essential to accept that most individual animals experience pain and do suffer, even if it is not the same sort of pain and suffering that humans, or even other nonhumans, including members of the same species, experience. Furthermore, when all individuals are admitted to the community of equals, their rights must be vigorously protected regardless of their cognitive skills or their capacities to experience pain and to suffer.

Deepening ethology also means that we need to bond with the animals we study and even name them (Davis and Balfour 1992). Many individual animals come to trust us, and we should not breach this trust. As L. E. Johnson has argued, "Certainly it seems like a dirty double-cross to enter into a relationship of trust and affection with any creature that can enter into such a relationship, and then to be a party to its premeditated and premature destruction" (1991, 122).

Let me emphasize once again that studying nonhuman animals is a privilege that must not be abused. We must take this privilege seriously. Although some believe that naming animals is a bad idea because named animals will be treated differently—usually less objectively—than numbered animals, others believe just the opposite, that naming animals is permissible and even expected when working closely with at least certain species, especially with the same individuals over long periods of time. As Manes notes, "If the world of our meaningful relationships is measured by the things we call by name, then our universe of meaning is rapidly shrinking. No culture has dispersed personal names as parsimoniously as ours . . . officially limiting personality to humans, . . . [and] animals have become increasingly nameless. Some*thing* not *somebody*" (1997, 155).

Early in her career the well-known primatologist Jane Goodall had trouble convincing reviewers of one of her early papers that naming the chimpanzees she studied should be allowed. She refused to make the changes they suggested, including dropping names and referring to the animals as "it" rather than "he" or "she," or "which" rather than "who," but her paper was published. It seems noteworthy that researchers working with nonhuman primates and some cetaceans usually name the animals they study; we read about Kanzi, Austin, Sherman, Koko, Phoenix, and Akeakamai and often see pictures of them with their proud human companions. We also read about Alex, an African gray parrot whom Irene Pepperberg has studied extensively. Yet most people do not seem to find naming these individuals to be objectionable. Is it because the animals who are named have been shown to have highly developed cognitive skills? Not necessarily, for these and other animals are often named before they are studied intensively. Or in the case of most nonhuman primates, is naming permissible because these individuals are more similar to humans than are members of other species? Why is naming a rat or a lizard or a spider more off-putting than naming a primate or a dolphin or a parrot? We need to know more about why this is so.

The context in which animals are used can also inform attitudes that people have even to individuals of the same species. For example, scientists also show different attitudes toward animals of the same species depending on whether they are encountered in the laboratory or at home; many scientists who name and praise the cognitive abilities of the companion animals with whom they share their home are likely to leave this attitude at home when they enter their laboratories to do research with other members of the same species. Based on a series of interviews with practicing scientists, Phillips relates that many of them construct a "distinct category of animal, the 'laboratory animal,' that contrasts with namable animals (e.g., pets) across every salient dimension. . . . [T]he cat or dog in the laboratory is perceived by researchers as ontologically different from the pet dog or cat at home" (1994, 119).

We must also pay attention to the oftentimes limited use, success, and even knowledge of animal models (see LaFollette and Shanks 1996; Shapiro 1997) and to the many successes of using nonanimal alternatives.

(It seems safe to say that most people would not venture to go to work if they had as little a chance of reaching their destination as some models have of helping humans along.) And we must not be afraid of what those successes may mean in the future—the reduction and then the abolition of animal use as models based on computer simulations or work on humans emerge superior.

Everyone must be concerned with the treatment of nonhuman animals, not only the rich and those with idle time on their hands. David Hardy has concluded that a detailed exploration of problems associated with animal well-being "must be consigned to those who have independent sources of wealth, no family obligations, and a lamented shortage of concrete worries" (1990, 11). I disagree. Everyone needs to be concerned with the treatment to which nonhumans are subjected. We must not only think of the animals when it is convenient for us to do so. Although the issues are at once difficult, frightening, and challenging, this does not mean they are impossible to deal with. Certainly, we cannot let the animals suffer due to our inability to come to terms with difficult issues.

We also need to teach our children well, for they are the custodians of the future. They will live and work in a world in which science increasingly will not be seen as a self-justifying activity but as another human institution whose claims on the public treasury must be defended. It is more important than ever for students to understand that to question science is not to be antiscience or anti-intellectual and that to ask how humans should interact with animals is not in itself to demand that humans never use animals. Questioning science will make for better, more responsible science, and questioning the ways in which humans use animals will make for more informed decisions about animal use. By making such decisions in an informed and responsible way, we can help to ensure that in the future we will not repeat the mistakes of the past, and that we will move toward a world in which humans and other animals may be able to share peaceably the resources of a finite planet.

We and the animals whom we use should be viewed as partners in a joint venture. We must broaden our taxonomic concerns, and funding must be made available for those who choose not to work on nonhuman primates. We must not be afraid of what broadening our taxonomic interests may bring concerning animal cognitive abilities and their ability to feel pain and to suffer. We must not continue to view animal suffering from afar, nor should we blind ourselves to the many ways in which we cause harm to the world around us. Chödrön argues that often "Our style is so ingrained that we can't hear when people try to tell us, either kindly or rudely, that maybe we're causing some harm by the way we are or the way we relate with others. We've become so used to the way we do things that somehow we think that others are used to it to" (1997, 33). Furthermore, Savage-Rumbaugh has stressed: "I believe it is time to change course. It is time to open our eyes, our ears, our minds, our hearts. It is time to *look* with a new and deeper vision, to *listen* with new and more sensitive ears. It is time to *learn* what animals are really saying to us and to each other" (1997, 68, my emphasis). These three Ls, looking, listening, and learning, should be used to motivate us to act on behalf of all animals.

Humans can no longer be at war with the rest of the world, and no one can be an island in this intimately connected universe. Nobel laureate geneticist Barbara McClintock claimed that we must have a feeling for the organisms with whom we are privileged to work (see Keller 1993). Thus, bonding with animals and calling animals by name are right-minded steps. It seems unnatural for humans to continue to resist developing bonds with the animals they study. By bonding with animals, one should not fear that the animals' points of view will be dismissed. In fact, bonding will result in a deeper examination and understanding of the animals' points of view, and this knowledge will inform further studies on the nature of human-animal interactions.

What I fear the most is that if we stall in our efforts to take animal use and abuse more seriously and fail to adopt extremely restrictive guidelines and laws, even more insurmountable and irreversible damage will result. Our collective regrets about what we failed to do for protecting animals' rights in the past will be moot. One way to begin is to expand the GAP and implement the GA/AP and admit all animals into the community of equals. It should be presupposed that at least some animal research and other activities that violate the rights of animals must not continue—the burden is on those who want to engage in these activities even if in the past they were acceptable.

My overall conclusion remains unchanged from that which I expressed in 1997 (Bekoff 1997a). Specifically, if we forget that humans and other animals are all part of the same world—the more-than-human world—and if we forget that humans and animals are deeply connected on many levels of interaction, when things go amiss in our interactions with animals, as they surely will, and animals are set apart from and inevitably below humans, I feel certain that we will miss the animals more than the animal survivors will miss us. The interconnectivity and spirit of the world will be lost forever, and these losses will make for a severely impoverished universe.

Notes

Parts of this essay were read at the conferences "Applied Ethics in Animal Research," Albuquerque, New Mexico (May 1997), and "Perspectives on Animal Consciousness," Wageningen, The Netherlands (July 1977). It contains some material from some of my previous essays (especially 1998b, with kind permission of Kluwer Academic Publishers), and acknowledgments to the numerous colleagues who have helped me along can be found in them. John Reed and Paul Moriarty discussed some aspects of this version with me. I especially want to thank my esteemed colleagues and close friends, Colin Allen and Dale Jamieson, for their extensive collaborative efforts.

1. See also Marler (1996) and Allen and Bekoff (1997), who come to this conclusion from different perspectives.
2. See Rachels (1990) for a discussion of species-neutral moral individualism and Bekoff and Gruen (1993) and Frey (1996) for further discussion.
3. See Bekoff (1991) and Bekoff and Elzanowski (1997) for further discussion.
4. Also see Weston (1994) concerning the notion of trans-human etiquettes.
5. Much of this section is taken from Bekoff (1998a), which discusses numerous aspects of cognitive ethology.
6. See Lehner (1996) for detailed discussions of ethological methods.
7. For discussion of this question with respect to fish, see Dionys de Leeuw (1996) and Verheijen and Flight (1997).
8. For further discussion and criticisms of ladder views, see Bekoff (1992), Crisp (1990), Fouts (1997), Pinker (1994), Sober (1997), and Verheijen and Flight (1997).

References

Abram, D. 1996. *The spell of the sensuous: Perception and language in a more-than-human world.* New York: Pantheon Books.

Allen, C., and M. Bekoff. 1995. Cognitive ethology and the intentionality of animal behaviour. *Mind and Language* 10:313–28.

———. 1997. *Species of mind: The philosophy and biology of cognitive ethology.* Cambridge, Mass.: MIT Press.

Baron-Cohen, S. 1995. *Mindblindness: An essay on autism and theory of mind.* Cambridge, Mass.: MIT Press.

Beck, B. B. 1982. Chimpocentrism: Bias in cognitive ethology. *Journal of Human Evolution* 11:3–17.

Bekoff, M. 1991. Animal ethics reconsidered. *Hastings Center Report,* September/October, p. 45.

———. 1992. What is a "scale of life?" *Environmental Values* 1:253–56.

———. 1993. Common sense, cognitive ethology, and evolution. In *The great ape project: Equality beyond humanity,* edited by P. Cavalieri and P. Singer, 102–8. London: Fourth Estate.

———. 1994. Cognitive ethology and the treatment of nonhuman animals: How matters of mind inform matters of welfare. *Animal Welfare* 3:75–96.

———. 1995a. Marking, trapping, and manipulating animals: Some methodological and ethical considerations. In *Wildlife mammals as research models: In the laboratory and field,* edited by K. L. Bayne and M. D. Kreger, 31–47. Greenbelt, Md.: Scientists Center for Animal Welfare.

———. 1995b. Naturalizing and individualizing animal well-being and animal minds: An ethologist's naiveté exposed? In *Wildlife conservation, zoos, and animal protection: A strategic analysis,* edited by A. Rowan, 63–129. Grafton, Mass.: Tufts Center for Animals and Public Policy.

———. 1995c. Play signals as punctuation: The structure of social play in canids. *Behaviour* 132:419–29.

———. 1995d. Vigilance, flock size, and flock geometry: Information gathering by western evening grosbeaks (*Aves, fringillidae*). *Ethology* 99:150–61.

———. 1996. Cognitive ethology, vigilance, information gathering, and representation: Who might know what and why? *Behavioural Processes* 35:225–37.

———. 1997a. Deep ethology. In *Intimate relationships, embracing the natural world,* edited by M. Tobias and K. Solisti, 35–44. Kosmos, Stuttgart: Beyond Words.

———. 1997b. "Do dogs ape" or "do apes dog?" Broadening and deepening cognitive ethology. *Animal Law* 3:13–23.

———. 1998a. Cognitive ethology: The comparative study of animal minds. In *Blackwell companion to cognitive science,* edited by W. Bechtel and G. Graham. Oxford: Blackwell.

———. 1998b. Deep ethology, animal rights, and the great ape/animal project: Resisting speciesism and expanding the community of equals. *Journal of Agricultural and Environmental Ethics* 10:269–96.

———. 2000a. *The smile of a dolphin: Remarkable accounts of animal emotions.* Washington, D.C.: Discovery Books.

———. 2000b. *Strolling with our kin.* New York: Lantern Books.

———, ed. 1998c. *An encyclopedia of animal rights and animal welfare.* Westport, Conn.: Greenwood Press.

Bekoff, M., and C. Allen. 1997. Cognitive ethology: Slayers, skeptics, and proponents. In *Anthropomorphism, anecdote, and animals,* edited by R. W. Mitchell, N. Thompson, and L. Miles, 313–34. Albany: State University of New York Press.

Bekoff, M., and A. Elzanowski. 1997. Collecting birds: The importance of moral debate. *Bird Conservation International* 7:357–61.

Bekoff, M., and L. Gruen. 1993. Animal welfare and individual characteristics: A conversation against speciesism. *Ethics and Behavior* 3:163–75.

Bekoff, M., and N. Hettinger. 1994. Animals, nature, and ethics. *Journal of Mammalogy* 75:219–23.

Bekoff, M., and D. Jamieson. 1991. Reflective ethology, applied philosophy, and the moral status of animals. *Perspectives in Ethology* 9:1–47.

———. 1996a. Ethics and the study of carnivores: Doing science while respecting animals. In *Carnivore behavior, ecology, and evolution,* edited by J. L. Gittleman, 2:15–45. Ithaca, N.Y.: Cornell University Press.

———. 1996b. *Readings in animal cognition.* Cambridge, Mass.: MIT Press.

Bentham, J. [1789] 1948. *An introduction to the principles of morals and legislation.* Reprint, New York: Hafner Press.

Bermond, B. 1997. The myth of animal suffering. In *Animal consciousness and animal ethics,* edited by M. Dol, S. Kasanmoentalib, S. Lijmbach, E. Rivas, and R. van den Bos, 125–43. Assen, The Netherlands: Van Gorcum.

Birke, L. 1997. Science and animals: Or why Cyril won't win the novel prize. *Animal Issues* 1:45–55.

Boone, J. A. 1954. *Kinship with all life.* New York: HarperCollins.

Boonin-Vail, D. 1994. Contractarianism gone wild: Carruthers and the moral status of animals. *Between the Species* 10:39–48.

Burghardt, G. M. 1997a. Amending Tinbergen: A fifth aim for ethology. In *Anthropomorphism, anecdote, and animals,* edited by R. W. Mitchell, N. Thompson, and L. Miles, 254–76. Albany: State University of New York Press.

———. 1997b. Review of Cavalieri and Singer. *Society and Animals* 5:83–86.

Byrne, R. 1995. *The thinking ape: Evolutionary origins of intelligence.* New York: Oxford University Press.

Carruthers, P. 1989. Brute experience. *Journal of Philosophy* 86:258–69.

Cavalieri, P., and P. Singer, eds. *The great ape project: Equality beyond humanity.* London: Fourth Estate.

Cheney, D. L., and R. M. Seyfarth. 1990. *How monkeys see the world: Inside the mind of another species.* Chicago: University of Chicago Press.

Chödrön. 1997. *When things fall apart.* Boston: Shambhala.

Crisp, R. 1990. Evolution and psychological unity. In *Interpretation and explanation in the study of animal behavior.* Vol. 2, *Explanation, evolution, and adaptation,* edited by M. Bekoff and D. Jamieson, 394–413. Boulder, Colo.: Westview Press.

Darwin, C. [1871] 1936. *The descent of man and selection in relation to sex.* Reprint, New York: Random House.

Davis, H., and D. Balfour, eds. 1992. *The inevitable bond: Examining scientist-animal interactions.* New York: Cambridge University Press.

Dennett, D. C. 1995. *Darwin's dangerous idea: Evolution and the meanings of life.* New York: Simon and Schuster.

de Waal, F., and F. Aureli. 1996. Consolation, reconciliation, and a possible cognitive difference between macaques and chimpanzees. In *Reaching into thought: The minds of the great apes,* edited by A. R. Russon, K. A. Bard, and S. T. Parker, 80–110. New York: Cambridge University Press.

Dionys de Leeuw, A. 1996. Contemplating the interests of fish. *Environmental Ethics* 18:373–90.

Fouts, R. 1997. *Next of kin: What chimpanzees have taught me about who we are.* New York: William Morrow.

Frey, R. G. 1996. Medicine, animal experimentation, and the moral problem of unfortunate humans. In *Scientific innovation, philosophy, and public policy,* edited by E. F. Paul, F. D. Miller Jr., and J. Paul, 181–211. New York: Cambridge University Press.

Gluck, J. P. 1997. Learning to see the animals again. In *Ethics in practice: An anthology,* edited by H. LaFollette, 160–67. Cambridge, Mass.: Blackwell.

Griffin, D. R. 1981. *The question of animal awareness: Evolutionary continuity of mental experience.* 2nd ed. New York: Rockefeller University Press.

———. 1992. *Animal minds.* Chicago: University of Chicago Press.

Hardy, D. T. 1990. *America's new extremists: What you need to know about the animal rights movement.* Washington, D.C.: Washington Legal Foundation.

Howard, W. E. 1994. An ecologist's view of animal rights. *The American Biology Teacher* 56:202–5.

Jamieson, D. 1985. Experimenting on animals: A reconsideration. *Between the Species* 1:4–11.

Jamieson, D., and M. Bekoff. 1992. Carruthers on nonconscious experience. *Analysis* 52:23–28.

———. 1993. On aims and methods of cognitive ethology. *Philosophy of Science Association* 2:110–24.

Johnson, E. 1991. Carruthers on consciousness and moral status. *Between the Species* 7:190–93.

Johnson, L. E. 1991. *A morally deep world: An essay on moral significance and environmental ethics.* New York: Cambridge University Press.

Keller, E. F. 1993. *A feeling for the organism.* San Francisco: W. H. Freeman.

Kellert, S. R. 1997. *Kinship to mastery: Biophilia in human evolution and development.* Washington, D.C.: Island Press.

LaFollette, H., and N. Shanks. 1996. *Brute science: Dilemmas of animal experimentation.* New York: Routledge.

Leahy, M. 1996. Brute equivocation. In *The liberation debate: Rights at issue,* edited by M. Leahy and D. Cohn-Sherbok, 188–204. New York: Routledge.

Lehner, P. N. 1996. *Handbook of ethological methods.* 2nd ed. New York: Cambridge University Press.

Lynch, J.J. 1994. Is animal pain conscious? *Between the Species* 10:1–7.

Manes, C. 1997. *Other creations: Rediscovering the spirituality of animals.* New York: Doubleday.

Marler, P. 1996. Social cognition: Are primates smarter than birds? In *Current ornithology,* edited by V. Nolan Jr. and E. D. Ketterson, 13:1–32. New York: Plenum Press.

Mason, W. A. 1979. Environmental models and mental modes: Representational processes in the great apes. In *The great apes,* edited by D. A. Hamburg and E. R. McGown, 277–93. Menlo Park, Calif.: Benjamin/Cummins.

Mench, J. 1996. Animal research arouses passion, sparks debate. *Forum for Applied Research and Public Policy* (spring): 5–15.

Mukerjee, M. 1997. Trends in animal research. *Scientific American* 276:86–93.

Phillips, M. T. 1994. Proper names and the social construction of biography: The negative case of laboratory animals. *Qualitative Sociology* 17:119–42.

Pinker, S. 1994. *The language instinct.* New York: William Morrow.

Pluhar, E. B. 1993a. Arguing away suffering: The neo-Cartesian revival. *Between the Species* 9:27–41.

———. 1993b. Reply. *Between the Species* 9:77–82.

———. 1995. *Beyond prejudice: The moral significance of human and nonhuman animals.* Durham, N.C.: Duke University Press.

Poole, T. B. 1997. Happy animals make good science. *Laboratory Animals* 31:116–24.

Premack, D. 1988. Does the chimpanzee have a theory of mind? Revisited. In *Machiavellian intelligence,* edited by R. Byrne and A. Whiten, 160–79. New York: Oxford University Press.

Rachels, J. 1990. *Created from animals: The moral implications of Darwinism.* New York: Oxford University Press.

Rennie, J. 1997. The animal question. *Scientific American* 276:4.

Ristau, C. 1991a. Aspects of the cognitive ethology of an injury-feigning bird, the piping plovers. In *Cognitive ethology: The minds of other animals: Essays in honor of Donald R. Griffin,* edited by C. Ristau, 91–126. Hillsdale, N.J.: Lawrence Erlbaum.

———, ed. 1991b. *Cognitive ethology: The minds of other animals: Essays in honor of Donald R. Griffin.* Hillsdale, N.J.: Lawrence Erlbaum.

Robinson, W. S. 1997. Some nonhuman animals can have pains in a morally relevant sense. *Biology and Philosophy* 12:51–71.

Rollin, B. E. 1992. *Animal rights and human morality.* 2nd ed. Buffalo, N.Y.: Prometheus Books.

Savage-Rumbaugh, E. S. 1997. Why are we afraid of apes with language? In *Origin and evolution of intelligence,* edited by A. B. Scheibel and J. W. Schopf, 43–69. Sudbury, Mass.: Jones and Bartlett.

Savage-Rumbaugh, E. S., S. L. Williams, T. Furuichi, and T. Kano. 1996. Language perceived: Paniscus branches out. In *Great ape societies,* edited by W. McGrew, L. F. Marchant, and T. Nishida, 173–84. New York: Cambridge University Press.

Shapiro, K. J. 1997. *Animal models of human psychology: Critique of science, ethics, and policy.* Kirkland, Wash.: Hogrefe and Huber.

Shepard, P. 1996. *Traces of an omnivore.* Washington, D.C.: Island Press.

Sober, E. 1997. Morgan's canon. In *The evolution of mind,* edited by D. Cummins and C. Allen. New York: Oxford University Press.

Taylor, P. W. 1986. *Respect for nature: A theory of environmental ethics.* Princeton, N.J.: Princeton University Press.

Thomas, E. M. 1996. *Certain poor shepherds: A Christmas tale.* New York: Simon and Schuster.

Tinbergen, N. 1963. On aims and methods of ethology. *Zeitschrift für Tierpsychologie* 20:410–33.

———. 1989. *The study of instinct.* 1951. Reprint, New York: Oxford University Press.

Tomasello, M., and J. Call. 1997. *Primate cognition.* New York: Oxford University Press.

Verheijen, F. J., and W. G. Flight. 1997. Decapitation and brining: Experimental tests show that after these commercial methods for slaughtering eel *Anguilla anguilla (L.),* death is not instantaneous. *Aquaculture Research* 28:361–66.

Wemelsfelder, F. 1997a. Investigating the animal's point of view: An enquiry into a subject-based method of measurement in the field of animal welfare. In *Animal consciousness and animal ethics,* edited by M. Dol, S. Kasanmoentalib, S. Lijmbach, E. Rivas, and R. van den Bos, 73–89. Assen, The Netherlands: Van Gorcum.

———. 1997b. The scientific study of subjective concepts in models of animal welfare. *Applied Animal Behaviour Science* 53:75–88.

Weston, A. 1994. *Back to earth: Tomorrow's environmentalism.* Philadelphia: Temple University Press.

White, R. J. 1990. Letter. *Hastings Center Report,* November/December, p. 43.

Whiten, A., and R. Ham. 1992. On the nature and evolution of imitation in the animal kingdom: Reappraisal of a century of research. *Advances in the Study of Behavior* 21:239–83.

Ethics,
Animal Welfare, and ACUCs

Bernard E. Rollin

Abstract: *In this chapter, Bernard E. Rollin, one of the architects of the animal-welfare legislation in the United States, addresses the criticisms of those who believe that Animal Care and Use Committee (ACUC) oversight is an example of the fox guarding the henhouse. He argues that the perspective of science as a value-free activity and the denial that animals have a subjective life deserving of consideration have been significant barriers to ethical evolution that have now substantially been overcome. He shows that the traditional concern for purposeful cruelty of animals, which has historically motivated advocates, must now be replaced by a recognition that the methods of modern agriculture and biomedical and behavioral research result in substantial unintended suffering. Dealing with this issue requires a different stance than one of unavoidable suffering in pursuit of scientific goals.*

Rollin takes on the issue of whether animals have rights and offers the challenging position that the encoding of animal-welfare laws that require scientists to consider some animal pain and suffering is de facto evidence of an acknowledgment that animals do have a form of rights. Specifically, he argues that those rights flow from an understanding and appreciation of their evolved natures. Readers should compare this perspective with those put forth by Frey and Biller-Andorno.

However, Rollin also acknowledges that current protections are inadequate in a number of important areas. He criticizes gaps in species protection that leave rats, mice, and birds uncovered, as well as the oversight exemption given to agricultural research. In addition, he makes unique recommendations about public involvement in oversight committees and concern for nonexperimentally produced suffering.

In a 1988 paper and in a recent book (Finsen and Finsen 1994), philosopher Lawrence Finsen, a thoughtful and rational animal advocate, has derided the value of Animal Care and Use Committees (ACUCs), the conceptual and operational foundation of federal legislation protecting research animals, as "a new set of clothes for the emperor"; in other words, as a sham. According to Finsen's view, a perspective also shared by many others in the animal-rights and welfare movement, such committees generally act as a rubber stamp for research use of animals, since "the membership of these committees are [*sic*] typically selected because they are favorably disposed toward research. Yet that these review committees exist allows researchers to claim that protocols are all carefully reviewed to assure that no 'unnecessary' suffering occurs in laboratories" (Finsen and Finsen 1994, 260). Many others have echoed this concern, claiming that current oversight mechanisms amount to little more than the fox watching the chickens. Because both major 1985 laws protecting laboratory animals—the National Institutes of Health (NIH) Reauthorization Act and the amendments to the Animal Welfare Act—do amount to what one Australian sociologist aptly described as "enforced self regulation," Finsen's criticism is a serious one (Rollin 1991). In its most interesting and strongest version, it suggests that current laws cannot move animal treatment beyond the status quo and certainly cannot effect major protective changes. If this is indeed the case, it is cause for despair both on the part of those of us who devised this system as the best hope for advancing the proper treatment of animals in research (Rollin 1993b) and on the part of the general public, whose deep concern for research animals led to the passage of these laws even in the face of significant opposition from the research community.

In what follows, I shall argue both on conceptual and empirical grounds that Finsen's view is misguided and that ACUCs must inevitably move beyond the status quo and have indeed done so already. Finsen's concern probably best applies to the rate at which change occurs, not the fact of such change.

Implicit in Finsen's charge is an "us" versus "them" gestalt. Ordinary people who put their faith in the laboratory-animal laws presumably possess one moral view of animals, and researchers possess a radically different and incommensurable one. Thus, self-regulation must inevitably fail. As I shall show, in one sense Finsen is historically correct. Where he goes astray, I think, is in misunderstanding the role of the laws. In my view, these laws must inevitably serve to bridge the gulf, not to perpetuate it. They exist not to divide scientists from the public on the moral status of animals but rather to assure, as Pogo says, that "from a moral point of view, they is us."

In order to understand this claim, we must first articulate and understand some of the deep philosophical preconceptions from which science has operated for much of the twentieth century.

Although twentieth-century science has tended, quite intentionally, to separate itself from philosophical concerns, it is patent that no area of human activity can avoid making philosophical commitments; all disciplines must rest on concepts and assumptions taken for granted by practitioners of the discipline. Twentieth-century science, too, has its philosophy, although that philosophy is typically invisible to its practitioners, who tend to see the assumptions of science not as debatable philosophical precepts but as self-evident truths. Thus, the philosophical assumptions made by science include an aversion to philosophical examination of these assumptions, and partly for that reason they have tended to harden into an ideology virtually universally pervasive among scientists, which I have elsewhere termed the common sense of science, for it is to science what ordinary common sense is to daily life (Rollin 1989).

One major component of scientific common sense directly relevant to the issue of animal use in biomedicine is the belief that science is value free, ought to make no valuational commitments, and thus, a fortiori, has no truck with ethics. This notion, like many other components of scientific common sense, is rooted in the logical positivism of the early twentieth century, which stressed the need for objectivity, empiricism, and verificationism in science: because value claims in general and ethical claims in particular are not subject to empirical test and verification, they have no place in science. They are at best, to scientific ideology, emotional predilections and cannot be dealt with objectively. It is for this reason that otherwise cool and rational scientists often become every bit as emotional regarding such ethical issues as animal use as their opponents—their training and ideology have led them to the view that ethical issues are in fact nothing but emotional issues, where rational thought has no place, and they thus believe that battles are won by manipulating emotions and tugging heartstrings. The possibility of a rational ethic on anything is instinctively seen as an oxymoron or solecism.

Even the most cursory examination of scientific writings of all sorts patently buttresses my claim that at least until very recently scientists distanced themselves from ethics. Keeton and Gould, for example, in their widely used college freshman biology text, remark that "science cannot make value judgements . . . and cannot make moral judgements" (1986, 6). In the same vein, Mader, in her basic biology text, asserts that "science does not make ethical or moral decisions" (1987, 15). James Wyngaarden, former director of the NIH, declared that all the flap about genetic engineering was misdirected because "science should not be hampered by ethical judgements" ("Director Addresses" 1989, 8). In 1988 Richard Marocco, a psychological researcher at the University of Oregon, responded to critics of animal research by asserting that their concern was "not an intellectual concern—it's an emotional, an ethical one, and a moral one" (U. 1989, 2), as if ethical concerns were not suited for rational adjudication. An American Veterinary Medical Association (AVMA) editorial in the 1960s stresses that the use of animals in research is a "scientific" and not a moral issue.

Such an approach is, for example, trumpeted in the passage chartering the AVMA Animal Welfare Committee, which proclaims that "AVMA positions should be concerned primarily with the scientific aspects of the medical well-being of animals, rather than the philosophical or moral aspects" (AVMA 1982). It is echoed in the preface to the Council for Agricultural Science and Technology (CAST) Report, appropriately entitled *Scientific Aspects of the Welfare of Food Animals,* which "focuses primarily on the welfare of food animals. . . . The Animal Rights issue has to do with human ethics and not with science; consequently it is not addressed at length" (CAST 1981). Amazingly enough, as the CAST statement implies, even the concept of animal welfare is seen as value free. Indeed, if one searches the major scientific literature on animal welfare, one will find a radical distancing of the "scientific" concept of welfare from any ethical dimensions. Suggestions such as that of Cambridge ethologist Donald Broom (1988), who defines welfare as adaptation to or coping with the environment, eschew any commitment to moral values. Even the most sophisticated views of animal welfare, such as those offered by Marian Dawkins (1980) and Ian Duncan (1981), views committed to the reality of animal subjective experience, presuppose the conceptual separability of animal-welfare science and ethical judgments, with scientists supplying value-free data and society making ethical judgments.

The idea that science is value free is reflected in the teaching of science, where science educators typically make such a doctrine explicit or implicitly communicate it by their failure to discuss ethical issues occasioned by the material they are teaching. Leading scientists in public pronouncements promulgate the value-free view of scientific inquiry. An excellent example of this may be found in a classic PBS television documentary dealing with the Manhattan Project, the development of the atomic bomb during World War II. When queried as to their ethical stance on the development of the bomb, most of the scientists replied that they left such considerations to the politicians, since ethics is not in the purview of scientists. Given this ideology, it is not surprising that the research community is often uncomfortable and inarticulate in its discussion of burgeoning social-ethical concerns (consider, for example, the issues of genetic engineering and cloning).

As Jay Katz at Yale University has documented, the medical-research community failed to see the moral issues in the use of human subjects for research until forced to address them by the threat of legislation in the 1960s, when Congress became concerned about the proliferation of revelations that the research community was using subjects without informed consent (Katz 1989). The Mississippi syphilis studies, the injection of live cancer cells into cancer patients by Brooklyn physicians, and the use of psychoactive drugs on mental patients and soldiers provided vivid examples to Congress. Willowbrook and recent revelations regarding radiation studies conducted by the U.S. Department of Energy provide later examples.

Few scientific journals, conferences, or courses proactively discuss the ethical questions engendered by their activities. (I was astounded to learn from the Office of Technology Assessment of Congress in 1987 that my 1985 paper on the ethical issues raised in the genetic engineering of animals [Rollin 1986] was the only paper on the subject published in the United States.) Nor is it surprising that leading scientists often say silly things about moral questions—hence, Donald Kennedy's incredible non sequitur about critics of animal research, namely, that "antivivisection was one of the policies of the Hitler regime" (Holden 1989). (As one of my students remarked, Hitler also had a mustache.)

In sum, scientific ideology denied the relevance of morality to science in general and, a fortiori, to the issue of animal use in research. Perhaps the most extreme example of this position can be found in a letter to the Hastings Center Register by researcher Robert White, professor of surgery at Case Western Reserve, who in a letter entitled "Animal Ethics?" condemned the journal for even treating the use of animals in research as a moral issue:

> I write in reference to the Special Supplement entitled "Animals, Science, and Ethics," which appeared in the May/June (1990) issue of the *Hastings Center Report*.
>
> I am extremely disappointed in this particular series of articles, which, quite frankly, has no right to be published as part of the Report.
>
> Animal usage is not a moral or ethical issue and elevating the problem of animal rights to such a plane is a disservice to medical research and the farm and dairy industry. (White 1990, 43)

The result of this component of scientific ideology was of course a blindness to moral issues occasioned by animal research. Typical of this reaction was the response I received in the early 1980s at a formal dinner from a prominent medical researcher who asked me to outline my concerns about animal research. I indicated that my chief concern was that many scientists do not even admit that there are any moral issues in this area. He then accused me and "all others like you" of attempting to "lay their trip on everyone else." "Morality," he told me, was "a matter of taste in a free country." To attempt to pass legislation restricting animal research was "fantastic" and "totalitarian." I was entitled to my opinion, he admitted, but ought not to try to impose it on him. He continued in that vein at length and wound down only when I pointed out to him that absence of constraints on the use of animals in research meant that he "imposing his trip on me," which was equally unacceptable, by his reasoning, in a democracy. In the same vein, no less a scientific institution than the *New England Journal of Medicine* accused me of being an ally of the Nazis and an apologist for lab trashers for suggesting the legislated constraints on animal use that have become law (Visscher 1982).

A second component of scientific ideology that insulates researchers from what ordinary common sense views as morally problematic in animal research has been the denial of scientific reality to animal consciousness, thought, and feeling. This component of scientific ideology holds that one cannot legitimately assert that animals are conscious in the sense of enjoying subjective experiences, feeling pain, fear, anxiety, loneliness, boredom, joy, happiness, pleasure, and the other noxious and positive mental states that figure so significantly in our moral concern for humans. This skepticism about attributing thought to animals enjoys a long history and was most famously promulgated by Descartes, who declared that animals were simply machines, driven by clockwork. Such a position, of course, provided justification for experiments in the burgeoning science of physiology in Descartes's time, which required dissection of living animals without anesthesia. While many philosophers and scientists and even theologians hotly contested Descartes's position, agnosticism about animal minds resurfaced in the early twentieth century, receiving succor from non-Cartesian sources. I mentioned earlier that the positivism that shaped scientific ideology denied the validity of talking about ethics, since moral claims were not verifiable. The same positivistic tendency nurtured the development of psychological behaviorism, which denied the ability to study the mind and consciousness and affirmed that only overt behavior was open to scientific inquiry. This methodological aversion to treating mental states as real had enormous influence, shaping the thinking of psychologists, zoologists, biomedical scientists, and even European ethologists who otherwise rejected behaviorism.

It is clear that the denial of mentation to animals did have untoward moral consequences in science. Scientific books and papers routinely stopped short of attributing felt pain, fear, and so on to animals, and any such extrapolations beyond overt behavior were seen as pernicious "anthropomorphism," despite the fact that much animal research presupposed that animals could feel pain. Although all analgesics in the United States were routinely tested on laboratory animals, these animals virtually never received analgesics in the course of research, and one searched in vain for literature on laboratory animal analgesia. When, serving on an American Association for Advancement of Laboratory Animal Science (AAALAS) panel in 1980, I challenged my copanelists, five prominent laboratory veterinarians, to tell me the analgesic of choice for a rat used in a crush experiment, none could tell me. "After all," some said, "we don't even know rats feel pain." Incredibly, the first conference on animal pain ever held in the United States was only convened in 1983, and even then, it dealt almost exclusively with the machinery or "plumbing" of pain, ignoring the subjective and morally relevant aspects (Kitchell and Erickson 1983). The scientific literature never discussed suffering in animals, and in its zeal to avoid "unverifiable" talk about mental states such as fear, anxiety, loneliness, and boredom, the research community talked in blanket terms of mechanical physiological "stress responses," which tended to be simplistically defined in terms of Cannon's alarm reaction for short-term stress or Selye's activation of the pituitary-adrenal axis for long-term stress.

These two components of scientific ideology taken together represented a formidable barrier to the development of scientific thinking about the moral issues associated with animal research. Qua scientist, in the role of scientist, scientists would don scientific common sense along with their lab coats and doff them when they went home to their role as ordinary people. Scientific common sense and ordinary common sense were, as psychologists put it, very well compartmentalized.

On one occasion, I was having dinner with a group of senior veterinary scientists, and the conversation turned to scientific ideology's disavowal of our ability to talk meaningfully about animal consciousness, thought, and awareness. One man, a famous dairy scientist, became quite heated. "It's absurd to deny animal consciousness," he exclaimed. "My dog thinks, makes decisions and plans, etc., etc." He then proceeded to exemplify these with the kinds of anecdotes we all invoke in such common-sense discussions. When he stopped, I turned to him and asked, "How about your dairy cows?" "Beg pardon?" he said. "Your dairy cows," I repeated. "Do they have conscious awareness and thought?" "Of course not," he snapped, then proceeded to redden as he realized the clash between ideology and common sense and what a strange universe this would be if the only conscious beings were humans and perhaps his dog.

In general, then, science and common sense coexisted like multiple personalities or Siamese twins, in the same body.

These are the major barriers I and my scientist and veterinarian colleagues faced when we convened in 1976 to draft the concept of legislation for laboratory animals. As we dialogued, it became clear to us that the aforementioned ideological barriers were the major impediments to improving the situation of laboratory animals. Somehow, we felt, one had to break down the separation between ordinary common sense (and its ordinary consensus morality) and scientific ideology's denial of morality in science and consciousness in animals. In essence, the task was to force scientists to reappropriate ordinary common sense and ordinary morality.

The way to do this, we felt, was to get scientists to think in different tracks, to force them to negotiate, as it were, on a moral playing field. (In my own work at Colorado State University in the 1970s, I had used this strategy to good effect in changing many invasive uses of animals, such as multiple survival surgery.) Common morality about animals had developed in the 1960s and 1970s to a point well beyond the traditional concern for prohibiting deliberate, intentional, willful, sadistic cruelty and promoting patronizing "kindness." Society has begun to realize that the overwhelming majority of the suffering that animals undergo at human hands is not the result of deliberate cruelty (Rollin 1993a, 1993b). Certainly well under 1 percent of animal suffering at human hands arises from sadism. The bulk of animal suffering in the mid–twentieth century in fact stemmed from the rise of intensive agriculture and on a lesser scale, the rise of significant use of animals in research and testing. These uses in turn stem from decent and laudable motives—producing cheap and plentiful food, curing disease, advancing knowledge, and protecting the public. Thus a new social ethic was needed "beyond cruelty," to express the public concern for suffering that is not the result of intentional cruelty. While the public continued to want the benefits of animal use, it also wanted to see the animals used treated fairly, not suffering, and indeed, so far as possible, happy.

In traditional extensive agriculture, the major use of animals in society, this demand was met because animals lived in environments for which they were biologically evolved. Producers did well if and only if animals did well. If one hurt an animal, one hurt oneself. Thus, in such a "contract" situation, cruelty was the major concern. Now, however, our uses of animals in science and agriculture do not assure their well-being. So society demands other ways of restoring the fairness to animal use. To do so, it in essence has done two things. First, it seeks new laws to govern animal use. Equally important, it takes as the moral justification for these laws a concept borrowed from human ethics, the concept of rights.

As Plato pointed out, one cannot teach ethics; one can only remind. That is, ethical progress must draw on ideas already in people's ethic, albeit perhaps unnoticed. When society sought a new language for expressing its desire for limitations on the use of animals, it naturally looked to an existing and, over the last forty years, much-emphasized idea in human ethics—the idea of rights. Rights are a protection for fundamental interests of human beings against encroachment by the majority or the common good. Thus we as a society spend a great deal of money to protect unpopular speakers even if no one wants to hear them, because we consider speech to be an essential feature of a human being's nature. It is the same with protection against torture, holding on to one's property, believing what one wishes, and so on. Very simply, society is asking that when we use animals we protect the fundamental aspects of their natures and the interests flowing therefrom and encode such protection in law. Hence our legislation mandated control of pain, suffering, and distress, as well as accommodation of research animals that suits their natures. (Congress changed this part of the legislation to cover only exercise for dogs and accommodations for primates that augment their psychological well-being.) Tellingly, both the head of the U.S. Department of Agriculture (USDA) division charged with enforcing the 1985 laws and two heads of the Institute for Laboratory Animal Resources (ILAR) have recently agreed with me that the new laws do (conceptually) codify certain basic rights for animals.

Our idea therefore was to remind scientists of what they as ordinary citizens must adhere to in the treatment of animals—very likely an idea that they accept when not wearing their lab coats (80 percent of the general public and 90 percent of the eight thousand or so Western ranchers I have addressed believe that animals have rights in the sense just described). Laws, to paraphrase Plato, are social ethics "written large." So if the law states that animals suffer and that such suffering must count and be dealt with in scientific deliberations, and that animal-care committees must work on the playing field of these assumptions, scientific ideology or common sense is suppressed in favor of ordinary common sense and consensus social morality. If federal law states that animals feel pain and suffer, scientific ideology cannot respond with agnosticism about animal consciousness. If federal law states that it is morally wrong to ignore animal suffering, scientific ideology cannot say that science is morally neutral and value free.

A wonderful anecdote illustrates this point well. When the law passed, Robert Rissler, a veterinarian who was then head of the Animal and Plant Health Inspection Service (APHIS), the branch of the USDA charged with writing regulations to interpret the law, was understandably concerned about the meaning of "psychological well-being of primates." Seeking counsel, he approached the primatology division of the American Psychological Association. "Don't worry," he was told, "There is no such thing." "Well, there will be after January 1, 1987, whether you guys help me or not," Rissler replied.

Thus the laws force the scientific community, through the vehicles of the ACUCs, to engage animal research in terms of common morality— to "recollect"—and also to reappropriate common sense about animal consciousness and pain. For members of ACUCs and indirectly for the scientists they represent, notice is served that scientific ideology is no longer adequate to determine the rules for animal research. (I have seen this work well in human subject committees. In the twenty-one years I have served on ours, I have seen committee sensitivity move from a low point of "What is wrong with using your students for nutrition experiments" to a point of sensitivity that prohibits researchers from bothering people who don't return questionnaires.)

This, then, is the conceptual part of our response to Finsen. Insofar as the 1985 laws force the thought of scientists into ethical channels and channels that recognize the reality of animal pain, suffering, and distress, they must change the way scientists think and act. For this reason alone, Finsen's claim that science's use of animals is essentially untouched by these laws must be false. Members of ACUCs start to think differently and through their decisions and dialogue spread this thinking to their colleagues.

Perhaps what Finsen means is that change has not yet taken place in animal care and use to any significant degree. That is an empirical question and would itself represent a strong objection to the current regime. If it is going to take a hundred years to get scientists to think differently and act differently, the current laws are inadequate for now, whatever their eventual outcome may be.

I do not think that this version of Finsen's claim is correct either. As I haved served on an ACUC since the late 1970s and a human-subjects committee since the mid-1970s, convinced my institution to review protocols meaningfully since 1980, successfully defended the concept of ACUCs to the U.S. Congress and the Dutch, Australian, and South African governments, and consulted for many ACUCs across the United States, I know that these laws do in fact work to improve the situation of animals. It is true that they follow no predictable timetable, because different institutions, like different individuals, vary in their cultures, personalities, receptivity, and so on. With the advent of the Internet, however, insights are quickly shared and more homogeneity in ethical response will inevitably emerge.

The following list cites some examples of where animals are immeasurably better off by virtue of ACUC activity in accordance with the laws:

1. First and foremost, the recognition and control of animal pain has grown rapidly in theory and in practice. Some examples will illustrate this.

    The reappropriation of common sense is well illustrated in the report of the AVMA Panel on Pain and Suffering in Animals, promulgated directly after the federal laws were passed (AVMA 1987). The panel takes it for granted that animals feel pain and defends anthropomorphic extrapolation of pain from humans to animals on the reasonable grounds that we make similar extrapolations in the other direction in animal research. Furthermore, the same veterinarians mentioned previously who were agnostic about animal pain had no trouble funding analgesic regimens in 1986. So, they explained to me, they simply retrieved drug company data, since all analgesics are tested on animals. Thus did the law change people's perspective on the same data. Analgesia is now routinely used in research.

2. More and more, control of care has moved into the hands of experts (lab-animal veterinarians and central staff). In the past, everyone was an expert. As one prominent physiologist remarked to me, "M.D.-Ph.Ds used to say 'Hell, if we can take care of people, we can take care of animals,'" even though one could get an M.D.-Ph.D in an animal-using area of science, begin a major animal-research program, and never learn anything about the animals used except that they model a particular disease or syndrome, and even though the researcher was accountable to no one regarding animal care. Furthermore, care and husbandry was often provided by minimum-wage student employees, which in practice meant erratic feeding and watering, failure to detect disease, and failure to control other variables that led to bad animal care and bad scientific results. More and more, committees demand central care by trained personnel and demand that

researchers demonstrate some mastery of basic principles of research, for example, surgery. Furthermore, courses are being mandated for nascent researchers in various aspects of animal care and use.

3. Even though the laws do not require it, many committees have extended the law beyond its letter, for example, to concern about pain in invertebrates or about surgery and pain control in farm animals used for agriculture research.

4. Committees have undertaken or forced the undertaking of research to the benefit of animals. For example, at my own institution, the ACUC mandated that a toxicology researcher who never used anesthesia for his painful studies for fear of skewing his results look at a variety of anesthetic regimens to see if such skewing actually occurred. (He ended up finding a regimen that did not affect his data.) We also fund small research projects that aim to benefit the animals.

5. Euthanasia protocols are tightly controlled. In the past, researchers used (or tried to use) such methods as succinylcholine or magnesium sulfate. Now committees even worry about animals being euthanized in front of other animals.

6. Many researchers are now undertaking work in enriched environments for species such as mice and rabbits that are not legally mandated to have such environments (*Lab Animal* 1994; *Lab Animal* 1995).

7. Committees are forcing teachers to think through anew the invasive uses of animals in teaching. Correlatively, many committees have mandated alternatives for students who don't wish to hurt or kill an animal for their education.

These are just a few clear examples of changes that belie Finsen's thesis. Nonetheless, it does seem at times that committees do move very slowly. In my experience, a peculiar phenomenon, which I call a breakthrough experience, is required to help things move forward until a certain momentum is reached. I don't believe such breakthrough experiences can be orchestrated, but they do inevitably occur as people become increasingly sensitized.

For example, one researcher on our ACUC had tended to view the whole protocol-review activity as a "waste of time" because researchers "know what they are doing." However, when we reviewed a research proposal to drown pigs in order to study refloat time for human drowning victims, he was shocked out of his sanguine stance. Under his leadership, the committee rejected the proposal, and he pointed out that a simple alternative was available—to study gas formation in slaughterhouse gut. Ever since that event, he takes protocol review very seriously.

Having, I hope, cleared the laws of Finsen's charges, I want to conclude by pointing out some limitations and inadequacies in the new laws. Despite the salubrious developments I have sketched here, the new laws are by no means ideal or even totally adequate. First, they do not cover all animals used in research. Neither of the new laws applies to rats and mice, farm animals, or birds used in industry research because the current USDA interpretation of the Animal Welfare Act still does not consider these creatures to be animals, and the NIH law only applies to federal-grant recipients or NIH's own labs. Clearly, the Animal Welfare Act must be extended to cover all animals; a federal judge ruled that the exclusion by the USDA of rats, mice, birds, and farm animals used in biomedicine actually seems to go against the intent of the law.

In addition, these laws are currently restricted in their application either to warm-blooded animals (Animal Welfare Act) or to vertebrates (NIH law). These cut-off points are clearly arbitrary, and many committees have to their credit extended the application to such higher invertebrates as squids, where there are good scientific reasons to suspect the presence of thought and feeling. The scope of these laws should be statutarily expanded to include all animals where there are solid reasons to infer the presence of pain or consciousness.

Another marked inadequacy in these laws pertains to animals used in agricultural research. The Animal Welfare Act specifically excludes from its purview farm animals used in agricultural research; NIH policy, too, does not apply to farm animals used in agricultural research. Yet millions of farm animals are used in such research in ways that may be as invasive and occasion as much pain and suffering as biomedical research. Such agricultural projects may include surgery, deficient diets, food and water deprivation, total confinement, and induced disease, yet these animals enjoy no legal protection. Thus, suppose one has twin male lambs, one of which goes to an NIH-funded biomedical research project, the second to an agricultural research project. Both are to be castrated. The NIH lamb will get anesthesia and postsurgical analgesia and will be castrated under aseptic conditions. The agricultural lamb may have the testicles removed under field conditions in standard ways—which even include having them bitten off.

To their credit, many committees now apply NIH standards to all surgical procedures done on their campuses—even for agricultural research. Nonetheless, agricultural animals clearly need to be included under standards as rigorous as those governing the treatment of biomedical animals. Although the agricultural-research community has adopted voluntary guidelines for their research animals, not surprisingly, they are far too weak and have no enforcement structure to back them. Yet another criticism stems from the fact that researchers are on the honor system to do what they say they will do in their protocols.

The major criticism of these new laws, however, is that they don't go far enough. Some philosophers have made the case that ultimately there is no moral justification for invasively using animals in research at all. When researchers attempt to answer this sort of argument, they respond in cost-benefit terms—that the good to humans and animals coming out of research outweighs the cost in pain and suffering to the research animals. Leaving aside the cogency of this response, one can indeed acknowledge that this statement seems to capture the current state of social moral thought on this question. But if this is indeed the case, then it naturally follows that the only invasive research that ought to be pursued is research in which the benefit to humans or animals likely to emerge from the research outweighs the cost in suffering to the animals. I have elsewhere called this the utilitarian principle (Rollin 1993b).

This maxim suggests that much invasive research, which is aimed at "pure knowledge," should not be allowed. One standard researcher response to this principle is to invoke the serendipity argument. It is argued that although it may not appear that a particular piece of research will produce foreseeable benefit, one never knows what will arise adventitiously. The response to that is simple. By definition, one cannot plan for serendipity. Society does not fund a great deal of research for a wide variety of reasons. Much research is turned down by the granting agencies because it is perceived as poorly designed, less important than other things, and so on. If the serendipity argument were valid, one could not make such discriminations, and one would be logically compelled to fund everything.

Admittedly, such cost-benefit calculation is fraught with difficulties—how does one weigh one parameter against a disparate one? But the crucial point to remember is that we do currently make such cost-benefit decisions in a variety of areas, including research on humans. All that needs to be done is to export such calculations to the area of animal use. Certainly, there will be hard cases, but at least extreme cases will be clear. Invasive research aimed at developing a new weapon or a new nail polish, or at discovering knowledge of no clear benefit to humans or animals—for example, territorial aggression studies—would obviously not be permitted.

Clearly then, some mechanism needs to be developed that will exclude invasive research that produces no benefit but simply advances trivial knowledge or careers. Some types of psychological research, for example, are highly vulnerable to this criticism. The current mechanism of peer review, whereby experts in the field judge the value and fundability of research, plainly does not address these concerns. Researchers who throughout their careers have taken a particular sort of invasive animal use for granted in their field are not the best source for eliminating such a use from the field. A better alternative, perhaps, would be to allow local committees with greater representation from the citizenry at large to pass on the value of a piece of animal research. Society pays for animal research; researchers ought to be able to successfully defend to a set of citizens their need to spend public money to hurt animals. Such an approach works for our justice system; perhaps researchers need to convince something comparable to a jury of their need to hurt animals for the sake of research.

Thus I would argue that local committees should also be charged with deciding whether a piece of research ought to be done and that such committees be made up of a majority of nonscientists representing the public in general.

The final major area of deficiency stems from the fact that the new laws focus on pain and suffering growing out of research use and only begin to touch on other forms of deprivation, growing out of husbandry and housing, exclusively in the case of dogs and primates. I would argue that moral concern, when applied to research animals, demands another principle, which I have elsewhere called the rights principle (Rollin 1993b), which asserts that in the context of research, all research should be conducted in such a way as to maximize the animal's potential for living its life according to its nature or telos and that certain fundamental interests should be preserved regardless of considerations of cost. In other words, if we are embarking on a piece of research that meets the utilitarian principle, we by no means have carte blanche; we must attend to the animal's interests following from its nature—the interest in being free from pain, being housed and fed in accordance with its nature, the interests in exercise and companionship if it is a social animal, and so on. The animal used in research should thus be treated, in Kant's terminology, as an end in itself, not merely as a means or tool.

I would therefore argue that the laws should mandate the creation of husbandry and housing systems that allow the animals to live lives approximating that dictated by their telos so as to assure as much as possible their happiness, as well as the mitigation of their pain and suffering. Precedent for this already exists in the work done on enriched environment for zoo animals. It has been argued, in fact, that animals suffer more in virtue of how we keep them than from what we do to them. I believe that the social ethic will eventually move in the direction I have just sketched.

References

American Veterinary Medical Association (AVMA). 1982. *Executive Board Minutes,* 14–15 July, 11.

———. 1987. Panel report of the colloquium on recognition and alleviation of animal pain and distress. *Journal of the American Veterinary Association* 191(10):1186–92.

Broom, D. M. 1988. The scientific assessment of animal welfare. *Applied Animal Behavior Science* 20:5–19.

Council for Agricultural Science and Technology (CAST). 1981. *Scientific Aspects of the Welfare of Food Animals.* Report no. 91. November.

Dawkins, M. S. 1980. *Animal suffering: The science of animal welfare.* London: Chapman and Hall.

Director addresses health research. 1989. *Michigan State News,* 27 February.

Duncan, I. J. H. 1981. Animal rights—animal welfare: A scientist's assessment. *Poultry Science* 60:489–99.

Finsen, L. 1988. Institutional animal care and use committees: A new set of clothes for the emperor? *Journal of Medicine and Philosophy* 13:145–58.

Finsen, L., and S. Finsen. 1994. *The animal rights movement.* New York: Twayne.

Holden, C. 1989. Universities fight animal activists. *Science,* 6 January, 18.

Katz, J. 1989. The regulation of animal experimentation in the United States: A personal odyssey. *IRB* 9(1):1–6.

Keeton, W. T., and J. H. Gould. 1986. *Biological science.* New York: Norton.

Kitchell, R., and H. Erickson. 1983. *Animal pain: Perception and alleviation.* Bethesda, Md.: American Physiology Society.

Lab Animal. 1994. February. Vol. 22(2).

Lab Animal. 1995. April. Vol. 24(4).

Mader, S. S. 1987. *Biology: Evolution, diversity, and the environment.* Dubuque, Iowa: William C. Brown.

Rollin, B. E. 1986. "The Frankenstein thing": The moral impact of genetic engineering of agricultural animals on society and future science. In *Genetic engineering of animals: An agricultural prospective,* edited by J. W. Evans and A. Hollaender. New York: Plenum.

———. 1989. *The unheeded cry: Animal consciousness, animal pain, and science.* Oxford: Oxford University Press.

———. 1991. Federal laws and policies governing animal research: Their history, nature, and adequacy. In *Biomedical ethics 1990,* edited by R. Humber and R. Almeder. Clifton, N.J.: Humana Press.

———. 1993a. Animal production and the new social ethic for animals. In *Food animal well-being 1993: Conference proceedings and deliberations.* West Lafayette, Ind.: Purdue University Office of Agricultural Research Programs.

———. 1993b. *Animal rights and human morality.* Buffalo, N.Y.: Prometheus Books.

U. 1989. *National College News,* 27 February, 8.

Visscher, M. 1982. Review of animal rights and human morality. *New England Journal of Medicine,* 306:1303–4.

White, R. 1990. Animal ethics? *Hastings Center Report,* November/December, 43.

Ethical Themes of National Regulations Governing Animal Experiments

An International Perspective

F. Barbara Orlans

Abstract: *In this chapter, Barbara Orlans, an experimental physiologist and research ethicist, identifies the possible components of a centralized regulatory structure that on the one hand acknowledges the potential value of animal research and on the other sets out requirements for that justification. She identifies eight regulatory dimensions that range from offering minimal to extensive protections to animal-research participants. For example, at one end of the continuum is the requirement of the provision of only basic husbandry needs and the formal inspection of research facilities by a government agency. At the other end, researchers are required to specifically estimate and justify the potential harm of the intended experimental procedures on the animals. Orlans then reviews the international distribution of animal-research laws and the extent of the provision of protection. Although the data reveals that centralized protection is missing in certain parts of the world, it also clearly shows that increased public concern since the 1970s has generally resulted in more stringent protection. She further argues that increased regulation is in fact an ethical advance and stems from the activities of the animal-rights movement and from the scientific discoveries that have improved the awareness of the intellectual and emotional capabilities of animals. In addition, Orlans discusses a number of controversial topics, such as the use of animals in education, the claim that animal use is decreasing, and the failure of the U.S. Animal Welfare Act to include rats, mice, and birds under its umbrella of protection.*

This essay reports on worldwide progress in the enactment of national laws governing the humane use of laboratory animals in biomedical research, testing, and education. During the last one hundred years, national laws to improve the welfare of laboratory animals have become enacted in at least twenty-three countries. I identify here eight ethical themes for discussion: (1) simple provision of basic husbandry requirements and inspection of facilities; (2) control of animal pain and suffering; (3) critical review of proposed experimental protocols; (4) specification of investigator competency; (5) bans on certain invasive procedures, sources of animals, or use of certain species; (6) application of the Three R alternatives—to refine procedures, reduce animal use, or replace animal procedures with nonanimal use where possible; (7) use of ethical criteria for decision making; and (8) mandatory use of animal-harm scales that rank degrees of increasing ethical cost to the animal. Countries in which all eight themes are addressed have the highest standards of animal care and use.

History

In 1876 Britain became the first country in the world to enact national legislation requiring minimal standards of laboratory animal care and inspections of laboratories for compliance. Another sixty years passed before any other country followed suit, and ninety before the United States enacted its first law governing humane use of laboratory animals, in 1966. Almost all of this legislation came about largely as a result of public protests about the conditions of laboratories and of the pain and suffering of the animals.

Since the late nineteenth century, animal experimentation has become a major tool of science and has expanded enormously. Nowadays, there is virtually no human disease—either physical or mental—that has not been investigated by the use of nonhuman animals. Diseases such as cancer, infectious diseases, drug dependence, and anorexia are induced or modeled in animals to determine the course of the pathological condition, its possible treatment, and prevention. Other uses include the development of pharmacologic agents, safety testing of consumer products such as cosmetics and detergents, and the training of students in experimental methods. An immense commercial enterprise has been established to supply laboratories with mice, rats, hamsters, guinea pigs, rabbits, dogs, cats, pigs, primates, and other species.

Public attitudes toward animals have changed dramatically over recent years. Since about the 1970s, there has been an upsurge in public concern about the treatment of animals in research. It stems both from the influence of the animal-rights movement and from the scientific discoveries that have widened our appreciation of the capabilities and feelings of animals. This awareness parallels the passage of stronger laws to protect animals.

Laboratory-animal protection laws usually cover all vertebrate species (both warm-blooded mammals and cold-blooded fish, reptiles, and amphibia). National policies of the United Kingdom and Canada are exceptional in that they also protect octopus, an invertebrate species. It is an anomaly not found in other countries that the U.S. law excludes rats, mice, and birds, the most commonly used laboratory species.

Countries with and without Animal Protection Laws

By 2000 at least twenty-three countries worldwide had enacted laws requiring certain humane standards for experimenting on animals: Australia (some states), Austria, Belgium, Denmark, Finland, France, Germany, Greece, Iceland, Ireland, Italy, Luxembourg, the Netherlands, New Zealand, Norway, Poland, Portugal, Spain, Sweden, Switzerland, Taiwan, the United Kingdom, and the United States. Canada is unusual in having led the way for a number of reforms without establishing a national law. Recently, the European Union has been influential in bringing countries without laws (such as Spain and Portugal) into line with other European countries that have well-developed laws. Only one country officially bans all animal experiments—the small European principality of Liechtenstein.

Currently, no legislation on humane use of laboratory exists in South American, African, or many Asian countries. Guidelines and voluntary controls may sometimes exist, but in the absence of any enforcement provisions, the policies may exist on paper alone.

In 1985 the World Health Organization promulgated *Guiding Principles for Biomedical Research involving Animals,* guidelines designed to provide a framework within which specific legislative or regulatory systems could be built in any country, including less-developed countries. Voluntary acceptance of these modest standards is better than having no provisions at all.

Ethical Issues in Current Laws

The eight issues I have previously listed can be used for comparison among the nations. Within the sequence of this listing is a loose, overall historical pattern. The first enactment of laws in any country typically deals only with the first two topics, basic husbandry requirements and inspection of facilities and control of animal pain and suffering. Only later (in amendments to the law) are refinements addressed that illustrate the next four topics—critical review of protocols, specification of investigator competency, bans on certain activities, and the use of the Three R alternatives. The last two topics, which address the complicated issue of how to justify each specific protocol, represent the cutting edge of new legislation. As yet, they are only found in laws of the most progressive countries concerned with animal welfare.

Husbandry Standards,
Inspections, and Record Keeping
Husbandry and Inspections

A basic ethical concern requires that captive animals be housed and cared for humanely. Official government inspection of research facilities maintains standards of sanitation, provision of food and water, space allocation by species' needs, daily care, and other basic requirements. Usually only minimum husbandry standards are mandated, and the tendency has been for animal facilities to conform to the lowest acceptable standards rather than providing optimal housing.

Inspections by government officials are needed to establish compliance. The frequency and adequacy of inspections vary from country to country, as do the standards required. In the United States, inspections are carried out once per year at each of the approximately fifteen hundred facilities registered with the controlling governing agency. In some countries, inspections are so infrequent and inadequate that the law exists on paper alone.

Historically, standards for housing space have been inadequate. For many decades, laboratory animals (primates, dogs, cats, guinea pigs, and rabbits) have typically been singly caged in barren environments. Commonly, they could be seen huddled in a corner because they had nothing to do. They were thwarted in their behavior by the poverty of their environments and deprived of any social interaction with other animals or stimulus from their environment.

However, housing standards for captive animals have been gradually improving in some countries. Reform has been sparked not only by more sympathetic public attitudes to animals but by research demonstrating that poor housing conditions cause stress to the animal, which can confound the experimental results obtained. Also, research has demonstrated that abnormal, stereotypic behaviors (such as pacing, cage biting, etc.) of laboratory, zoo, and farm animals do not occur if the animals are housed in enriched environments—ones as close as possible to those conditions experienced by free-living animals.

In the United States, Congress enacted an amendment to the Animal Welfare Act in 1985 that requires promotion of the "psychological well-being" of primates. This legal provision sparked new funding for environmental enrichment studies and has been profoundly effective in improving the housing conditions of primates. There is a trend toward increased space allocation, group-housing animals of similar species, and the addition of branches, toys, and exercise apparatus to the cages where appropriate. European countries and Australia have been in the forefront of enriching the housing of many common laboratory species, not only primates but also dogs, cats, rabbits, guinea pigs, and rats.

Record Keeping

Public reporting of the numbers and species of animals used is a basic requirement of effective oversight and accountability of animal experiments. The rationale is that the public has a right to know what is happening in this socially controversial area of harming animals for human good. Table 1 reports on nineteen geographic areas that account for 35,865,000 animals annually. Worldwide, estimates of the total number of animals used range as high as fifty to a hundred million annually, since many animals are uncounted.

TABLE 1

Number of Laboratory Animals
Used in Research, By Country

Country	Number of Animals Used
United States (1998)	12,138,000*
Japan (1995)	9,415,000**
United Kingdom (1998)	2,666,000
France (1997)	2,609,000
Canada (1998)	1,766,000
Germany (1998)	1,496,000
Australia (New South Wales, South Australia, Victoria, Tasmania, Western Australia) (1996/7)	1,141,000
Italy (1996)	1,094,000
Netherlands (1998)	674,000
Norway (1997)	630,000
Spain (1996)	507,000
Switzerland (1998)	492,000
Denmark (1997)	380,000
New Zealand (1999)	249,000
Sweden (1997)	267,000
Finland (1998)	195,000
Ireland (1996)	77,000
Portugal (1996)	50,000
Greece (1996)	19,000
Total	35,865,000

NOTE: Numbers of animals are given to the nearest thousand and represent official national statistics, except for the United States. Figures in parentheses indicate year of count. Figures may notbe directly comparable from country to country because of different criteria for counting (in the United Kingdom, for instance, procedures are counted rather than number of animals), and species covered by the law vary (e.g., Canada and some others include fish whereas others do not).

*The U.S. Department of Agriculture counts only about 10 percent of all animals used in experimentation. The most used species—rats, mice, and birds—are not protected under the relevant legislation and are therefore not counted. In 1998 the number of animals officially counted was 1,213,814. For this table, this figure has been multiplied by ten to allow for the uncounted species and to achieve approximate comparabilty with data from other countries.

**Data on "used animals" from the Japanese Association for Laboratory Animal Science.

It is unclear whether the total number of animals used worldwide is declining, as animal advocates hope, or increasing. A few countries, including the Netherlands, have reported a decline. In the United States, very probably the largest user of animals worldwide, inadequate data make trends impossible to assess. Within the estimated 10 percent of animals used that are counted, there has been a decline. On the other hand, biomedical research is increasing at a phenomenal rate and so, too, must be the numbers of animals used. In fiscal year 1999, the U.S. Congress appropriated a record $15.6 billion for the National Institutes of Health (NIH), a 15 percent increase over the preceding year. About half of this funding is spent on animal research. It is the stated intent of Congress to double NIH research funding from 1999 through 2004, which will further increase the number of animals used. A major new use of research animals is genetic engineering (mainly on mice), and the impact of this new direction will surely increase the number of this species used for research.

Controls on Animal Pain and Suffering

National laws also require that every effort be made to reduce or eliminate pain and suffering that result from an experimental procedure. Anesthetics, analgesics, and postoperative care should be used wherever needed, and animals in extreme pain should be put to death. It is generally considered a matter of plain humanity that the degree of animal pain and suffering be minimized. Indeed, it is a moral imperative.

But such provisions have not necessarily come with the first enactment of a national law. For instance, in the United States, animal pain was not addressed in 1966 when the law was first passed. Indeed, at that time, whether animals actually perceived pain was widely doubted. Not until 1976 was the Animal Welfare Act, the federal law governing laboratory animals, amended to require for the first time the use of anesthetics and analgesics. As a result, research on animal pain and its alleviation accelerated. Textbooks devoted to the physiology and relief of animal pain were published, new anesthetics and analgesics were developed, and postsurgical care became an important topic. Great progress has been made, and by now it is well recognized in national policies throughout the world that vertebrate animals do indeed feel pain.

Methods of killing of animals represent another aspect of control of pain. In the 1980s, the American Veterinary Medical Association established standards for recommended euthanasia practices to ensure that methods used are as rapid and painless as possible. These standards, which are now law in the United States, have been repeatedly updated, and other countries have adopted similar standards.

Critical Review of Protocols

Not all countries include legal provisions for review of investigators' proposed protocols. Sometimes investigators are subjected to either no formal procedure for protocol review or review only by their peers within their own discipline (for instance, in departmental review at a university or pharmaceutical company).

Nonetheless, there has been considerable growth in the establishment of oversight review committees that function as gatekeepers for approval of proposed experiments. These committees are variously called Animal Care and Use Committees (ACUCs) or Ethical Committees. Canada was the first country to establish an institutional committee review mechanism in 1984, and several other countries have now followed suit. Among these are (in chronological order) Sweden, the United States, Australia, New Zealand, the Netherlands, and the United Kingdom.

The ethical rationale behind such review is that investigators should be accountable in what they do, not only to their peers but also to the public. These committees may be institutional or regional. They typically operate with considerable autonomy, being only loosely regulated by national bodies. The resulting framework is often characterized as "enforced self-regulation."

The composition of these committees varies among countries, but most include representation of several viewpoints. Committee membership typically includes animal researchers, veterinarians, and lay (nonscientist) members of the public. Representatives from the animal-protection movement should be included because this is the constituency most concerned about humane standards. Experience has shown that to avoid rubber stamping, committee membership should not be dominated by animal researchers, and the chair should be an independent person and not an animal researcher.

The value of public representation on these committees is well established, and most committees would benefit with increased public representation. The New Zealand 1988 *Guidelines for Institutional Animal Ethics*

Committees states: "[L]ay people do more than add credibility: they provide valuable perspectives and reflect the concerns of the public." They are "vital to effective functioning" of local animal care committees and their input "opens the 'closed doors' to overcome the criticism of an internal peer review system" (National Animal Ethics Advisory Committee 1988, 16–17).

The purpose of oversight committee review is to ensure compliance with established standards of care and use by modifying (to improve the animal's welfare) or disapproving proposed projects. This does not necessarily mean that ethical debate that questions the fundamental justification of a project occurs. Indeed, most commonly there is an assumption of fundamental justification of a project. Thus there is room for considerable improvement in the level of debate within most of these oversight committees. I discuss this further in the section "Ethical Criteria for Decision Making."

Specification of Investigator Competency

The question concerning investigator competency is "What training is required before a person is allowed to conduct *any* animal experiment?" Untrained persons are likely to inflict greater harm on an animal than trained persons attempting the same procedure and furthermore are unlikely to produce experimental results that are of scientific value. So the benefits are less and the harms greater. Establishment of competency standards is important, yet only a few countries have adequately addressed these issues.

A few national laws have established specific educational standards for investigator competency. For instance, the 1972 Animal Protection Act of Germany states that experiments that could inflict pain, suffering, or harm to an animal "may be performed only by persons who have completed a college education in the field of veterinary medicine or medicine, and who have the required professional knowledge, and by persons who have completed an education in biology at colleges or state scientific institutes, insofar as these persons have the required professional knowledge." Similar provisions exist in Finland and the Netherlands.

In both the Netherlands and the United Kingdom, a mandatory full-time training course in humane experimental techniques and animal anesthesia is required before an investigator qualifies to conduct animal experiments. The course in the Netherlands is full-time for three weeks. In many countries, however, standards are permissive. The laws rather loosely require that an animal experimenter must be "qualified" or "trained." But if no educational standards are specified and no specific training mandated, the chances of effectiveness are slim.

In addition to controls over qualifications for persons working in animal-research facilities, several countries place controls over what is permitted by beginning biology students in early stages of their education. Switzerland permits almost no experiments involving animal harm in primary or secondary schools or at undergraduate college levels. The 1987 revised *Australian Code of Practice for the Care and Use of Animals for Scientific Purposes* includes a special section (7.3) on the use of animals in teaching. It bans the following activities: surgical procedures other than normal animal husbandry operations, induction of infectious diseases, production of nutritional deficiencies giving rise to distress, exposure to distressful stimuli, and administration of toxins.

Historically, a real problem existed in U.S. junior and senior high schools in the 1960s to early 1980s. Youths from age eleven to seventeen sought to impress judges of science-fair competitions by attempting highly invasive experiments on live animals. Often the students conducted these experiments in their homes, and supervision was absent or cursory. Extreme animal suffering occurred. Typical were high school student projects of attempted mammalian surgery, blinding, injection of lethal substances, and starving animals to death. Because the public protested strongly about these abuses, improvements have been made. But still today there is inadequate control over the use of animals in junior and senior high school education in the United States, as well as insufficient encouragement to use non-harmful alternatives. Federal laws do not exist. Unsatisfactory 1995 guidelines (which are voluntary and unenforceable) of the National Association of Biology Teachers include no provisions to ban the infliction of animal pain or suffering on sentient creatures and encourage dissection. Further reforms are still urgently needed.

As for the use of animals in U.S. colleges, there has been limited progress. The 1985 amendment to the Animal Welfare Act required for the first time that oversight committees review the use of animals in undergraduate college courses at some (but by no means all) tertiary educational institutions. Previously, decisions on what invasive animal experiments were to be included in student courses rested almost exclusively with the teacher. With this low level of review, some highly invasive experiments, such as brain ablation of mice or administration of electric shocks were encountered. After 1985 review committees began to disapprove many uses of animals in education. But by no means is this situation under control. Rats, mice, and birds, the species most used in college classes, are not covered under the Animal Welfare Act. Thus, a number of colleges fall outside the law, so that much of the use of animals in U.S. biology classes is unregulated.

There are a number of ethical rationales for prohibiting students from harming or killing animals: (1) nonpainful, nonharmful animal projects, nonharmful human studies, and other projects that carry no ethical burden are readily available that are equally or more instructive; (2) because projects at this educational level are primarily demonstrations of known facts, they lack the major ethical justification for harming animals that is based on the reasonable likelihood of obtaining significant, original knowledge; (3) unskilled students are likely to inflict greater harm than trained researchers; and (4) allowing emotionally immature youth to harm animals under the guise of education desensitizes students' feeling of empathy with animals. It can be argued that these points apply not only to primary and secondary school students but also to undergraduate college-level students. It is usually not until graduate school that a student makes a serious career commitment, and even then, not all careers in the biological sciences require expertise in animal experimentation techniques.

Bans on Certain Activities

Experimental procedures that cause intense and prolonged animal suffering have been the focus of the greatest public protest and demands for prohibition. Even if useful scientific results might be obtained, the lack of justification holds.

Some success in banning such activities has been achieved. A 1986 amendment to a German law, the Animal Protection Act, forbids experimentation on animals for development and testing of weapons, as well as the testing of tobacco products, washing powders, and cosmetics. The Netherlands and the United Kingdom also ban the use of animals for cosmetic testing. Recently, the British government announced its commitment to stop licensing any further testing of tobacco or alcohol products on animals. Indeed, in the whole field of animal testing, with the bans on the notorious LD50 test (the lethal dose that painfully kills 50 percent of the animals) and the Draize eye irritancy test (which can cause blindness in rabbits), considerable progress has been made.

Recently, three European countries (the Netherlands, Switzerland, and the United Kingdom) have banned the use of the ascites method of monoclonal antibody production. This procedure, used on mice, causes considerable suffering, including respiratory distress, circulatory shock, difficulty walking, anorexia, and other disabilities. It is estimated that in the United States up to one million animals a year are killed using this experimental method, but efforts to ban it in the United States have failed. In 1997 the American Anti-Vivisection Society launched a national campaign to ban this procedure because in vitro alternatives are available. However, the NIH did not approve the ban, and the U.S. Department of Agriculture, the federal agency that administers the Animal Welfare Act, was not involved because mice are not covered under this law.

Another issue, apart from the experimental procedure, is the source of the subject animal. There are three potential sources: former animal pets, either stolen for research or abandoned by their owners; free-living wild animals; or purpose-bred animals (those specifically raised by commercial breeders for research). All sources have come under criticism (antivivisectionists object to every source), but most criticism has focused on the use of one-time companion animals and on the capture of wild animals, especially nonhuman primates.

Of the three possible sources, the use of purpose-bred animals is preferred. The ethical reasoning is that purpose-bred animals are likely to suffer less; they do not have to make a stressful transition from a free life to a life in captivity. Purpose-bred animals know no other life than living in confined quarters; they have been singly caged all their lives with little or no opportunity to make decisions for themselves over what exercise they take, what they eat, whom they spend time with, and so on. But former pets and free-living wild animals are different; they have usually lived rich social lives where they were accustomed to expressing their own free will. To lose this freedom can be traumatic. The period of transition can cause considerable suffering, including the stresses that come with transportation (sometimes for thousands of miles, as with some nonhuman primates, and which can result in death), close confinement, and social and other forms of deprivation.

In addition, the experimental results from purpose-bred animals are more reliable because, unlike former pets and wild animals, their genetic and health backgrounds are known. This reduces the number of variables that can confound experimental results.

Efforts for reform in Europe have been largely successful in stopping the use of former pets for research. For instance, a 1997 law in the Netherlands, the Experiments on Animals Act, requires that animals used for research must be purpose-bred. The United Kingdom has banned since 1998 the use of chimpanzees (whatever their source), and there are constraints on the use of any species of wild-caught primates. Captive-bred alternative animals must be used there. The only exception is baboons, which cannot be bred in captivity. Such reforms have yet to come to the United States, which has no constraints over the use of wild-caught animals, and where approximately two thousand chimpanzees are housed in research laboratories. About half the number of dogs and cats used for research in the United States are former pets taken from shelters, and the other half are purpose-bred.

Three R Alternatives

The Three R principles (refine, reduce, replace), first enunciated by Russell and Burch in 1959, state that experimental procedures should be refined to lessen the degree of pain or distress, that the numbers of animals used should be reduced consistent with sound methodological design, and where possible, that nonanimal methods should be used in preference to those that do use animals. Legal mandates requiring the Three Rs facilitate the acceptance of these concepts by investigators and oversight reviewers. The countries that specifically address all Three Rs in their legislation include the United States, the Netherlands, Sweden, Switzerland, and New Zealand.

The Three R principles are increasingly becoming accepted worldwide by both the humane and scientific communities. Although antivivisectionists focus on replacement alternatives exclusively, others believe that incremental improvements in laboratory animal welfare are best achieved at this time by pursuing all Three Rs.

Promising advancements can be made in refining experimental methods by improving anesthetic and other pain-relieving regimens, using humane experimental end points, and employing only rapid and painless methods of euthanasia. To a lesser extent, reductions in numbers are feasible through the better use of statistics in methodological design. Replacement alternatives may not be applicable, but increasingly, nonanimal alternatives are being developed, especially in animal testing and in teaching biology to students. Evidence of the increasing role of all Three Rs is indicated by the success of the continuing series of World Congresses on Alternatives and Animal Use in the Life Sciences, influential international meetings that attract participants from about forty countries.

Although the concept of the Three Rs is now fairly well accepted on a universal basis as an ideal, it has proved very difficult to persuade regulatory bodies to stop requiring safety tests that involve use of whole animals before a new product can be approved. Although validated nonanimal tests are available in many cases, the regulatory bodies continue to mandate whole-animal testing. The nonanimal tests are thereby unreasonably being held to a much higher standard of validation than animal tests.

The evaluation of progress in implementing the Three Rs is a new topic and is in its infancy; most countries do not have adequate data for analysis. However, the Netherlands provides a unique model. Analysis of official data shows a significant decline in the percentage of total experiments that involve severe animal pain, from 29.3 percent in 1984 to 18.8 percent in 1997 (Orlans 2000). In addition, over the same period, the number of animals used has dropped by about half: in 1984 the total was 1,242,285 and in 1997 it was 618,432 (Orlans 2000). So there has been not only a reduction in numbers but evidence that refinements and replacements are being applied. In addition to this, since 1989 research establishments in the Netherlands report to the government whether new alternatives to animal experiments have been introduced in their laboratories.

Ethical Criteria for Decision Making

In general, existing laws do not address the fundamental ethical question, "Should this particular animal experiment be done at all?" The usual presumption of the law is that animal experimentation is justified and that proposed projects should be approved so long as the individual investigator believes that useful scientific knowledge might be gained. Indeed, oversight committees tend to approve almost everything that investigators propose, even highly invasive procedures on primates. Although some projects are modified (typically by application of a refinement), rarely is any proposal totally disapproved. It is thus a step forward when national policies specifically acknowledge that ethical decisions are involved in assessing the justification of an animal experiment, giving credence to the possibility that a proposed work is not justified.

Several countries have taken the lead in requiring a cost-benefit analysis that links animal pain (and other harms) to the scientific worthiness and social significance of the experiment's purpose. Typical is the *Australian Code of Practice,* which states: "Studies using animals may be performed only after a decision has been made that they are justified, weighing the scientific or educational value of the study against the potential effects on the welfare of the animals" (Australian Government Publishing Service 1990, 6). Similar provisions are found in national policies of the United Kingdom, Germany, the Netherlands, and Canada.

The concept of making a cost-benefit analysis sounds reasonable but is difficult to apply because the costs and benefits are incommensurable. Almost all the harms fall on the animals and all the benefits on humans. Nevertheless, the cost-benefit view has gained considerable acceptance as a tool for clarifying ethical choices. Acknowledgment that ethical issues are involved is found in the 1997 Dutch law, which recognizes "intrinsic value of animal life"—a forceful statement in this context of laboratory animals.

Use of Animal-Harm Scales

An important issue on the cutting edge of new reforms in national laws is the requirement to assess and rank the sum total of animal harms for any particular procedure. The ranking systems are variously called severity banding, invasiveness, or more colloquially and inaccurately, pain scales. First mandated in the Netherlands in 1979, such systems are now found in other countries (in chronological order, the United Kingdom, Finland, Canada, Switzerland, and New Zealand). This spread attests to the usefulness of these schemes. Pressure exists in the United States and other countries to adopt similar systems.

According to these systems, the degree of pain or distress is ranked according to a severity banding of either minor, moderate, or severe. For example, in the minor category are such procedures as biopsies or cannulating blood vessels; in the moderate category are major surgical procedures under general anesthesia and application of noxious stimuli from which the animal cannot escape; in the severe category are trauma infliction on conscious animals and cancer experiments with death as an end point. At some point (according to one's point of view), procedures become unethical because of the severity of animal pain.

Mandatory use of these ranking systems forces laboratory personnel to think carefully about the condition of the animal and its state of well-being or adversity throughout the experiment. It also encourages laboratory personnel to learn how to identify clinical signs of well-being and adversity.

In recent years, adoption of harm scales by various countries has acted as a significant stimulus to clinical investigations of animals in assessing signs of well-being and adverse states. A notable contribution that has attracted worldwide attention is that of Mellor and Reid (1994). Their categorization system, which represents a major step forward in assessing the condition of animals, has been adopted with minor modification as national policy in New Zealand and is the gold standard by which other harm rankings should be measured.

Summary

Laboratory animals are much benefited by enforcement of legally established standards for humane care and use. Nonetheless, an absence of laws in many countries where animal experimentation takes place needs to be corrected. New provisions along the lines of the topics discussed here are also needed, as is enforcement of many existing laws. It takes a great deal of effort to enact legal protections for animals, but the value of such laws has been indisputably established, as evidenced by the vast improvements that have come about in the standards of animal care and use found in today's laboratories compared with those of previous years. I also believe that improved conditions that serve to support the welfare of animals serve also to improve immeasurably the quality of the resulting science.

References

Australian Government Publishing Service. 1990. *Australian code of practice for the care and use of animals for scientific purposes*. Canberra: Australian Government Publishing Service.

Mellor, D. J., and C. S. W. Reid. 1994. Concepts of animal well-being and predicting the impact of procedures on experimental animals. In *Improving the well-being of animals in the research environment*, 3–18. Glen Osmond, South Australia: Australian and New Zealand Council for the Care of Animals in Research and Teaching.

National Animal Ethics Advisory Committee. 1988. *Guidelines for institutional animal ethics committees*. September 16–17. Wellington, New Zealand: National Animal Ethics Advisory Committee.

Orlans, F. B. 2000. Public policies on assessing and reporting degrees of animal harm: International perspectives. In *Progress in the reduction, refinement, and replacement of animal experimentation*, edited by M. Balls, A.-M. van Zeller, and M. E. Halder, 1075–1082. Amsterdam: Elsevier Science.

Russell, W. M. S., and R. L. Burch. 1959. *The principles of humane experimental technique*. London: Methuen. Reprinted 1992 by Universities Federation for Animal Welfare, 8 Hamilton Close, South Mimms, Potters Bar, Herts, UK ENG 3QD.

World Health Organization. 1985. *Guiding principles for biomedical research involving animals*. Geneva: Council for International Organizations of Medical Sciences.

The Importance of Nonstatistical Experimental Design in Refining Animal Experiments for Scientists, IACUCs, and Other Ethical Review Panels

David B. Morton

Abstract: *In 1959 William Russell and Rex Burch published a book entitled* The Principles of Humane Experimental Technique, *in which they argued that scientists have an ethical obligation to replace the use of animals in research whenever possible, reduce the numbers used to the fewest consistent with good science, and refine the experimental procedures such that the animals experience the minimal amount of pain and distress possible. These recommendations have come to be known collectively as the "Three Rs."*

In this chapter, David Morton, a laboratory animal veterinarian and ethicist, focuses on the process of experimental refinement. He, in essence, provides a practical handbook to researchers about the process, thereby providing guidance on the forms of suffering that concern Rollin. Morton suggests to the reader that little should be assumed and that what has often become standard practice in laboratories needs to be exposed to rigorous examination. Such examination begins with the scientific justification for conducting a particular experiment, an evaluation of the design and goals of the project, and an evaluation of the competencies of the research group. He emphasizes the importance of creating a process whereby all laboratory personnel have a vehicle for providing their ongoing assessment of the experiment. Morton's premise is that a great deal of expertise resides in the experience of animal breeders, veterinarians, and laboratory technicians. He specifically addresses crucial topics such as the use of pilot studies, historical control groups, death as an experimental end point, and long-term anesthesia. Of particular importance is his description of a data-collection scheme for tracking animal reactions to experimental procedures, which goes a long way to benefiting the science and the animal welfare of a specific project.

Beyond the practical and below the surface of these recommendations, Morton exemplifies an ethical approach to animal research that balances animal harms against scientific benefit. This approach leaves open the possibility that animal interests could at times override scientific progress.

Introduction: The Three Rs and the Limitations of Replacement

Refinement, one of the Three Rs, is perhaps the area wherein most progress can be made to reduce any animal suffering incurred in animal experimentation in the short term. Another "R," reduction, will reduce the number of animals used in research through good statistical experimental methods and design. In this essay, however, I wish to show how nonstatistical strategies can contribute markedly to minimizing pain and suffering in an experiment. The third "R," the replacement of animals in experiments, is a longer-term aim that requires considerable investment and basic animal biology data input before absolute animal alternatives can be made available to provide reliable scientific data. In the past twenty years or so, replacement has developed as a concept more in relation to toxicity testing and finding ways in which the necessary harms done to animals (i.e., the necessity to produce clinical signs of toxicity in order to determine the safety of a chemical) can be eliminated. In fact, less than 10 percent of animal experiments in the United Kingdom involve toxicity testing, and the use of replacement alternatives in the other 90 percent of experiments is less likely because whole-animal body systems are needed, since the purpose of the experiment is not simply to determine toxicity but to determine the function of a part of the body (e.g., the immune system or reproduction cycle) or to test the effectiveness of new medicines (e.g., antibiotics, antihypertensives, anxiolytics, analgesics, anesthetics) and so require whole animals to be fully evaluated. Nevertheless, even in these situations, in vitro methodology can provide important information for an investigator. It may seek to answer questions such as whether a chemical interacts with cell receptors or fails to interact, and at what level it is toxic.

Because the scientific questions to be answered in vitro differ from those in vivo, these alternatives have been called "adjuncts" and should precede in vivo work, in ways I discuss later in this essay.

Refinement

Refinement, I believe, is advanced as an attempt not only to reduce adverse states in animals but also to promote positive mental and physical states, and so I define it as

> Those methods that avoid, alleviate, or minimize the potential pain, distress, or other adverse effects suffered by the animals involved, or that enhance animal well-being.

The scientific paradigm is normally to change one variable at a time in some form of standard experimental protocol. Thus identifiable variables are limited as much as possible in any one experiment, so that only the specific variable of interest is examined. This objective is compromised not only by poor experimental design but also through inadvertent side effects that can occur as a result of compromised animal well-being. Unintended animal suffering deriving from physical damage to an animal, or physiological or psychological perturbances from normal, can materially affect the scientific data collected and so should be avoided as much as possible. Avoiding such confounding effects involves more than simply providing anesthetics and analgesics when animals may be in pain, or alleviating distress and fear in some way or another. Although these are important, additional strategies can be employed to reduce the level of animal suffering, notably statistical and nonstatistical design of experiments. Traditionally, research into the refinement of experiments has not been rigorously investigated because it is not seen to be part of the main line of research. Consequently, important questions such as how much pain should be caused to demonstrate effective analgesia or how long we should starve animals to make them work are often ignored, and instead, the standard or traditional level/time is used. If scientists wish to claim that they practice humane science, then they have to pay as much attention to avoiding unnecessary pain and distress to their animals as they do to their scientific objectives.

Humane science will influence both the welfare of the animals and the scientific quality, and it deserves more detailed consideration than it seems to have attracted to date. Moreover, attention to refinement will have a positive effect on the financial costs of the research because both experimental design and the welfare of the animals affect the variance in the scientific results. This means that well-designed experiments will be more likely to use fewer animals to achieve the scientific objective, which in turn will lead to decreased costs for upkeep, administration, record keeping, treatment, and so on. It is not unusual for important scientific discoveries to have to be validated through replication of the work in another laboratory, which will also add to the overall costs if the initial study was poorly carried out, especially if it was not accurately and fully written up. It is clearly important, therefore, that scientific protocols, including the precise experimental design, be recorded in detail in peer-reviewed journals (Morton 1992). In summary, factors leading to good animal welfare will usually also lead to humane, reliable, accurate, and economic science.

Literature Review and Choice of Model

One should establish three primary points before embarking on any experiment: Has the work been done before? Is there an alternative way of achieving the scientific objective without the use of sentient animals? and Is the scientific question worth answering in the light of the pain and suffering that will inevitably be caused to the animals during both their husbandry and the experimental procedures? (See table 1, nos. 5 to 9.) Table 1 details the various factors that should be considered and addressed before starting an experiment. Its theme will be picked up and illustrated throughout this essay.

It is important to review the literature before starting any work, not only focusing on whether the proposed approach to the scientific objective is appropriate but carrying out a critical analysis to determine whether the proposed model really will achieve the stated objectives. If there is a choice of models, then one must ascertain the advantages and limitations of each (see table 1, nos. 2, 3, 4, and 13). For example, is the model simply reproducing clinical signs of a disease or reflecting the actual cause of the disease such as the infective agent or the genetic defect? (Note: It is in this area that transgenic animals should improve the validity of certain animal models of human diseases.) Has the model been validated in any way? If so, how, and is it scientifically acceptable?

TABLE **1**

Animal Model Data.

1.   Model of:

Purpose of Research (Benefit)

2.   Purpose for which the model can be used (e.g., human disease, drug evaluation)
3.   Advantages and disadvantages (scientifically and in its relevance and application to human or animal disease or treatment)
4.   Clinical trials or medicine (drug) evaluation indication relevance or predictablitly of the model to its intended purpose (no. 2)

Animal Details

5.   Species of animal and experimental details (e.g., strain, inbred/outbred, sex, age, weight, health status, acclimation period, other)
6.   Husbandry special or critical requirements (e.g., caging or pen type, animal kept singly or in groups, diet, bedding, isolator or filration boxes, breeding details, other)
7.   Methodological details (e.g., equipment required, manual skills required, dosing and timings, timings of measurements, where advice can be obtained)
8.   Refinment aspects (e.g., humane endpoints, score sheet giving cardinal signs, useful tests, and any other information)
9.   Scoring of signs and severity grading (severity grading: give average and maximum expected)

Alternative Models and Staging Strategies

10.   Replacement (any alternative methods for all or part of the model available, e.g., pre-screening)
11.   Useful information to have before in vivo work begins
12.   Whether pilot studies are necessary for dose sighting and whether the doses chosen are the minimum necessary to obtain an indication of effectiveness before proceeding to more detailed trials, e.g., dose-response studies
13.   Success rate of the model (give mean and range and methods used to determine success)

14.   Statistics (how the data is to be handled; the best means of analysis, data information, etc.)
15.   If lethality is an end point, justify. If and LD50% dose is being obtained, justifiy the scientific necessity for a precise estimation in relation to the subsquent extrapolation to human or animal therapies
16.   Ethical commentary
17.   References
18.   Keywords for literature searches

The choice of species can be critical when the results are to be extrapolated to humans; those animal species with similar genetic predispositions or metabolic pathways or that show similar pathology to humans are likely to be better models. When the purpose of the work is specifically aimed at that species (e.g., veterinary, zoological, or wildlife research), then other considerations come in to play, such as the stresses involved in capture and confinement. It is also important to remember that there are strain or breed differences within a species that can be critical, as many researchers have shown (see table 1, nos. 5 and 6; Claassen 1994; Hendriksen and van der Gunn 1995). Thus both species and strain should be checked for scientific validity.

Other questions include whether adequate local facilities are available to produce the model; whether the necessary expertise is available to do this accurately and precisely; and whether the staff is able to care for the animal afterward and during its lifetime (see Smith and Boyd 1991, 141–46).

A Comment on "Inverse (Perverse?) Speciesism"

In some countries there is a level of speciesism, in that special justification is required for the use of some species (e.g., in the United Kingdom the use of dogs, cats, horses, and primates). This is not related to the biological needs of these animals or to their ability to feel pain more than other species. Rather it is an attempt to reflect the special public concern over the use of some species compared with others. Moreover, the "other species" are often eaten, regarded as pests, or are at least distasteful in some way or another even though they may also be kept as pets (e.g., mice, rats, gerbils, hamsters, guinea pigs, or rabbits). Surely, the rational choice of species is that group that is the least sentient consistent with sound science.

The Use of Anesthetics and Analgesics

Scientific procedures should be performed under anesthesia (general, regional, local, or with analgesia) unless either the pain or discomfort of the procedure is less than that caused by the anesthetic, or that the use of such regimens would be incompatible with the scientific objectives. The latter has to be shown and not assumed, and a scientific evaluation is the best way to do this. Too often, much is assumed, both to the detriment of animal well-being and possibly to the science because of the animals' responses to the stressors. By not researching this, however, it is never revealed. In addition, the type of anesthesia should be examined to ensure that it is suitable for both the species and the duration of the procedure and, of course, that the researchers are competent in its administration. It is possible to anesthetize animals for long periods (often at least twenty-four hours if not several days). For any experiment that may cause severe adverse effects during that time period, and for which anesthesia per se will have no effect on the experiment (e.g., some infection studies), serious consideration should be given to doing the work completely under anesthesia from which the animal never recovers (see table 1, no. 7).

The Importance of Competent and Skilled Investigators

It is important that any pain or suffering during an experiment should be kept to a minimum through the use of analgesics, and so on. It is here that competent and skillful researchers are so vital for the production of good science (see table 1, no. 7). Poorly carried out techniques can cause considerable and avoidable suffering, and it is in the interests of both science and animals that all scientific procedures be carried out in the most humane manner. An investigator should carefully learn a new technique that demands new skills before embarking on any experiment in which animals recover and data are recorded, even if more animals have to be used. Becoming competent in a technique involves watching those who are competent and learning the tricks of the trade. Depending on the technique to be learned, it may be appropriate to dissect dead animals to learn the anatomy and to understand the dangers of inaccuracy. One has to ensure that the apparatus or equipment to be used is compatible with the species, strain, sex, size, or age of the animals. For a surgical technique, the next stage may be to practice on recently killed animals as the tissues are warm, their texture is more lifelike, and hence the sensitivity of the procedure is more realistic for the operator. Finally, a surgical technique can then be applied to anesthetized animals that may be allowed to recover only if it has gone well; but if not, those animals should be killed. The importance of learning with a scientist or technician who is competent, skilled, and who keeps abreast of new developments cannot be overemphasized. Moreover, even simple techniques on conscious animals, such as handling, oral dosing, parenteral administration of substances, and removal of blood, if carried out badly can significantly elevate the level of animal pain and distress. This in turn will cause further problems for the next time as animals anticipate what is to about to happen to them based on their previous experiences. This in turn will impact on the science.

Statistical Advice

One should obtain advice, especially statistical advice, from a variety of sources before starting an experiment (see table 1, no. 14). I do not intend to discuss such advice in any depth here, as many standard texts and commentaries cover the various approaches (e.g., Chamove 1996; Erb 1990, 1996; Festing 1994; Khamis 1997; Mann, Crouse, and Prentice 1991; Mead 1988; van Zutphen, Baumans, and Beynen 1993). They include concepts such as avoiding using too few animals during an experiment as opposed to too many; the type of data to be collected and how the data should be analyzed; the value of pilot studies and prior information of expected variance; the calculation of the number of groups, the number of animals in each group, and increasing the power of an experiment through blocking; ongoing analysis during an experiment; and statistical instruments and tests.

The Limitations of Scientific Evidence, Extrapolation, and Animal Numbers

The purpose of an experiment also must be clearly defined, as should any application to which the results may be applied. Thus in safety testing for a new chemical entity, the necessity to obtain acute toxicity data to a probability of less than 0.001 is pointless when an extrapolation is to be made to another species, normally the human. Moreover, if a fudge factor, such as an additional safety margin of 100 times the toxic dose found in animals, will be used to calculate the hazard involved in any exposure for humans (e.g., for the transport of that substance), then the futility of such precise experimentation is highlighted even further. The question then has to be asked: What is going to be done with the results? If there is a clear difference in consequent actions based on the difference in probability estimates, then, providing that can be justified, it is acceptable. But if there is to be no difference in action between different statistical precisions, then the lowest acceptable precision that will lead to the use of the fewest animals should be used.

Another instance for which fewer animals could be used is when it has been decided that even if one animal in a group fails a test, then a specific action will take place. For example, if a substance is going to be labeled as toxic on the basis of one animal in a group of six reacting adversely, and if the first animal reacts adversely, then there is no point in continuing with the experiment and giving that substance to the remaining five animals. It would have provided valid scientific information, of course, but the practical applied outcome has already been decided, and so there is no point in continuing the study and causing pain and suffering to the remaining animals.

Protocols evaluating new treatments, such as the rodent protection test (RPT), can help determine, for example, effectiveness and therapeutic levels of antibiotics for use in humans. In the RPT the effectiveness of a novel compound to prevent or treat infection is investigated over a range of doses using standard microorganisms in an animal model (mouse or rat). If, compared with other similar drugs, the compound appears to have advantages (e.g., showing a better bacterial kill rate or less toxicity), then the information gained from the animal studies will be taken into account when phase one studies (healthy human volunteers) and phase two studies (human patients who can expect to benefit therapeutically and in whom effective dose levels will be determined) are undertaken. There is little need for high precision in the animal model species as opposed to the target species, and yet the approach of accepting a lower statistical probability ($p < 0.05$ compared with $p < 0.01$) may mean the difference between five and fifty animals in a group. This is especially important in the area of infection, where considerable suffering may occur in animals who are not protected from the disease. (One might even query why $p < 0.05$ is a scientifically justified probability level rather than $p < 0.06$ or $p < 0.04$.)

Pilot Studies, Dose Sighting, and Control Groups

Pilot studies can be useful to give some idea of the variance, likely adverse effects that may be encountered, and any practical local problems likely to be encountered during the main study. Pilot experiments help, therefore, to determine the appropriate number of animals that should be used in any one group, as well as how best to avoid and alleviate any adverse effects (see table 1, nos. 12 and 14). The use of pilot studies needs not be a waste of animals, because the results can be used for the main study, unless there are good proven scientific reasons for not doing so.

Historical data in this regard is very useful, particularly when considering both positive (e.g., a standard challenge of a known substance) and negative (e.g., giving vehicle alone) controls. The use of background data can influence the number of animals needed for a given study. In some instances it may not even be necessary to have a control group if the control "fact" is so well known and accepted (e.g., that total pancreatectomy leads to diabetes mellitus with a consequential rise in blood sugar and death in less than ten days). However, it may be advisable to carry out pilot studies when using a new model in one's own animal facility, because differences in strain of animal, diet, staff, bedding, cleaning materials, husbandry, and environment have all been shown to affect animals' responses in various ways (see, e.g., Claassen 1994; Morton 1995b; Wadham 1996). This was exemplified by a multicenter trial on the Fixed Dose Procedure carried out by van den Heuvel and coworkers (1990); they found that while the LD50 (the dose that kills 50 percent of the animals) of various chemicals was similarly ranked between laboratories, the actual LD50 dose varied considerably.

Several methods exist for determining the relevant dose(s) for animals in one's facility in pilot studies, but the up-and-down method deserves special mention (Bruce 1985). In this approach a single animal is given a test dose (e.g., from the literature), and its response is noted. Assuming that the first dose does not produce the anticipated response, a second animal is given a higher (normally threefold) or lower (normally by one-third) dose as appropriate. This procedure is then repeated on individual animals at varying dose intervals until a suitable dose is found. I have found this approach helpful for streptozotocin dosing to induce diabetes in various strains, weights, and ages of rats, as well as for other chemicals given to induce various disease states.

Advice from Other Scientists, Veterinarians, and Caretakers on Adverse Effects

When carrying out a novel scientific procedure for the first time, one will need a prediction of the type and incidence of any adverse effects on the animals; thereafter, the experience gained will help in future work using that or similar protocols. Although a thorough review of the literature is essential, sadly, the adverse effects on the animals encountered is rarely written up in a helpful manner for those using the model (see Morton 1992). Furthermore, the "tricks of the trade," so often vital for one to be able to repeat the work, are omitted, as are relevant details relating to the animal side of the work, such as those shown in table 1 (nos. 8 and 9). The use of score sheets (Morton 1994, 1995a,b, 1997a,c, 2000; Morton and Townsend 1994, 1995) help all those involved in the research program to recognize the cardinal clinical signs that animals show during a specific scientific procedure and can indicate consequent actions, notifications, and treatments to be taken when certain clinical signs are observed (e.g., humane end points). Animal caretakers, stock persons, and veterinarians who are familiar with the animals in the experiment are excellent guides in determining any adverse effects. They may not always be able to explain why an animal is "wrong" or "not right," but good carers are rarely mistaken. The score sheets, which begin to explain their insight, and simple tests, such as giving an analgesic and observing if animals change their behavior (e.g., by moving around more, eating and drinking), all add empirical evidence to identify those animals who were affected by that protocol and were in pain.

Order of Work

Scientists should have a clear idea of how the work is to be carried out, including the order of work, the interdependence between the proposed experiments, and any collaborative work with other research groups. Although some of the examples I give here may seem obvious, I ask the reader to consider their work carefully, since a well-structured research program from an animal-welfare viewpoint may not be quite the same as one from a scientific viewpoint, although the validity of any data would not be affected.

In Vitro to In Vivo

Research is sometimes aimed at investigating the interactions of substances (such as new drugs) in the whole body. If the in vitro work will yield information that shows the effectiveness of the substance, or that the viability of the cells/organisms has been compromised in some way, it may not be worth proceeding with the in vivo work. There has to be confidence that the in vitro work can be extended to the in vivo situation, and if so, the in vitro work should precede the in vivo experiments. For example, genetic modification of viruses may reduce their virulence in that they are no longer able to infect cells in vitro, and based on past experience, it is unlikely they will infect cells in vivo. Similarly, radiolabeling of bacterial toxins may remove their toxicity, which can be measured by a change in binding characteristics to monoclonal antibodies, or to cell surfaces in vitro, or to a change in toxicity to less-sentient animal forms such as invertebrates. Again, experience indicates that it may be unlikely that the modified toxin will be effective in vivo, and so these experiments need not be done. In cases of doubt, however, small pilot projects involving only two or three animals can be used (depending on the precision needed) to confirm such findings and so validate a predicted negative, especially when there is less likelihood of causing animals pain and suffering. When a positive harmful effect needs to be confirmed in vivo, then only one or two animals may be needed, as any toxicity will be likely to provide sufficient evidence (e.g., bacterial viability after a freeze-thaw cycle). Furthermore, in such cases where an adverse effect is predicted and needs to be confirmed, it may be possible to prevent the animal feeling any pain by carrying out the work under general or local anesthesia.

One can carry out specific tests to determine if an experiment is likely to work in vivo or whether it is unlikely to succeed for some reason. If a chemical is cytopathic (i.e., toxic to cells) in vitro at concentrations that were predicted to be useful in vivo, then the work should not proceed. This is the basis of the so-called prescreening tests used in the early stages of toxicity testing of new chemicals and medicines. An example of this might be a chemical that in solution at the desired concentration has a pH that is far from being biocompatible and will obviously cause tissue damage if given parenterally. It may then be given by some other route (e.g., enterally or at a lower concentration or in a different formulation). In all these cases consideration is given to carrying out in vitro work before whole-animal work in order to avoid, or alleviate, any adverse effects in animals.

Within In Vivo Work

It may be possible to stage work even within in vivo work so that serious adverse effects are avoided. For example, if one is looking at the survival of animals given a bowel transplant, those animals that fail are likely to die painfully from a peritonitis. One could wait for clinical signs of peritonitis to become apparent, but it would also be possible to avoid peritonitis altogether by trying to pick up signs of very early rejection, rather than bowel perforation (caused by rejection leading to death of intestinal wall cells and disintegration of the bowel). The early signs of transplant failure may be thrombosis of the vessels supplying the graft within a few hours of the end of surgery. By keeping the animal under terminal anesthesia for, say, twelve hours or so, one can monitor blood flow and the condition of the transplant (color, contractility, etc.), and if found to be deleteriously affected, the animal can be killed. If successful after twelve hours or so, the animal can be permitted to recover. Alternatively (and I would recommend), another animal can be permitted to go for twenty-four hours before being killed. In such a way animal suffering can be reduced considerably, and it may be possible to avoid serious adverse effects altogether.

Another approach is to always perform any critical pilot experiments before any controls or further experiments. Experiments should be made at the outset of any project that could provide key information on whether it is worth progressing down certain lines of experimental inquiry. For example, if an experiment is to determine the effect of a substance or specific cell or tissue on a physiological response, then a pilot study showing first that that response actually occurs should be undertaken. Only then should the main study and necessary controls, such as using saline, tissues or cells, sham operations, and so on, be conducted.

Mild to High Severity

Several examples can be instanced in which the adverse effects for an animal can first be shown to be effective or ineffective at a low level before going to higher levels of animal pain and suffering. Traditionally, this aspect of refinement has not been rigorously investigated because it has not been seen as part of the main research and the "standard/traditional" time or level normally used. In some psychological experiments animals may be starved for long periods in order to make them "work" in some way or another. If the animal can be motivated to work by reward, rather than by starvation or even aversion stimuli, then such a strategic progression of harm could reduce the amount of animal suffering without losing the scientific objective. If starvation is to be used to motivate the animal, then how long is sufficient: forty-eight hours, twenty-four hours, eighteen hours, twelve hours, six hours? And does it depend on the age of the animal, the species, the strain, the time of day, the diet, and so on? One must also consider whether the adverse physiological effects of such periods of starvation (and even dehydration) on the animals will affect their performance and hence the science. If aversion has to be used, then the same principle should apply; use the lowest level of aversion (e.g., electric current, sound level, learned helplessness) necessary to achieve the scientific goal. Testing of novel analgesics or determining neurocircuitry or neurotransmitters may be possible through causing low levels of painful stimuli rather than high ones. An analgesic ineffective at low levels of pain is unlikely to be effective at higher levels of pain; the neurobiology may change quantitatively but not qualitatively. In any event the purpose of the experiment has to be explicit, and justification must be made for the higher pain levels. Finally, the dose-sighting studies mentioned previously and practiced by the toxicologists in safety testing are not routine in other areas of science. A scientist may start a dose-response experiment at a range of dose levels only to find all the animals are unaffected, or worse still, that all the animals have died. Pilot studies are always worth doing in these sorts of cases because dose responses vary according to species, strain, diet, husbandry, environment, and so on (see van den Heuvel et al. 1990).

Other examples relate to trying to preserve an animal's quality of life and include only affecting the use of one of a paired organ rather than two. For example, use just one leg in fracture/arthritis/musculoskeletal research; for sense organs use one eye or one ear; in renal transplantation research, transplant a third kidney into an animal's groin, where it can easily be monitored, rather than removing an animal's own kidneys and replacing it with the "donor" kidney (this could also be an example of a staged approach if true dependence needed to be established). A similar approach can be adopted in the study of skin transplantation or burns research, where a minimum area should be transplanted or damaged; in nerve regeneration studies where nerve section or denervation is required, rather than incapacitating the whole leg by sectioning or crushing the nerve high up, use a low nerve section that will affect just one muscle in the distal part of the limb so that compensation is possible through the use of other muscles. In all these examples, if the purpose is to study quantitative aspects, the experiment can be staged so that small harms are first seen to be "effective" in whatever scientific sense is appropriate, and then the insult can be scaled up, providing the scientific justification is ethically acceptable.

Cancer Research

A set of excellent guidelines exists for the refinement of experiments involving cancer (Workman et al. 1998). The recent increase in cancer experimentation as a result of gene therapy work makes this an important area for refinement. The growth rate of tumors and their size should be kept to the minimum necessary to demonstrate effectiveness or ineffectiveness of novel therapeutic agents or regimes. Death can hardly ever be justified as an end point (Mellor and Morton 1997). Tumors should be placed whenever possible in sites where they are easy to palpate or monitor in some way and that cause minimum pain and discomfort. Animals should be monitored at least daily and more frequently if necessary (e.g., at times when the tumor ulceration is likely to occur or the animal may be found dead). This means that the growth characteristics of the tumor should be reliably evaluated and well known beforehand. Measures that indicate animal well-being in addition to tumor size should be sought as well, particularly if the tumor is deep, invasive, or internal and thus difficult to detect (e.g., in body cavities or skull). These may include the behavior of animals, especially at night when they should be maximally active, their coat quality, posture, appearance, body weight and body condition, and so forth. Finally, it may be that some therapies are directed specifically against tumors that are well established. In such cases the effectiveness of the therapeutic agent should first be demonstrated on small early tumors, as it is unlikely that more developed ones will be treatable if these are not affected.

The Medical Research Council (1998) has addressed implementation of humane end points in cancer research in humans; parts of this work are equally applicable to animal studies. Response to a potential therapy is measured along the following lines:

Complete response: the disappearance of all known malignant disease.

Partial remission: at least 50 percent reduction in the sum of the
products of the two largest perpendicular diameters of all measurable lesions. In addition, there can be no appearance of new
lesions or progression of any lesion.

No change/stable disease: a 50 percent decrease in total tumor size of
all measurable lesions cannot be established, nor has a 25 percent increase in the size of one or more measurable lesion been
demonstrated.

Progressive Disease: a 25 percent or more increase in the size of one
or more measurable lesion, or the appearance of new lesions.

Humane End Points

It is often possible, given sufficient familiarity and experience with an animal model or specific scientific procedure, that the adverse effects on animals can be predicted. These may include the clinical signs an animal
might show, the duration of those signs, the progression to other signs
(even death), the timing of such signs after a particular procedure has
been carried out, and the proportion of animals that may be affected.
Based on such predictions, actions can be taken to minimize animal suffering, depending on the scientific purpose as well as humane considerations. Certain levels of harm may be integral to the procedure and so unavoidable to achieve the scientific objective, but that may not always be
the case. There are several instances when animals should be killed before the planned end of an experiment. For example, when an animal in
pain and distress has deviated physiologically or psychologically so far
from normal that it has lost its scientific utility, it should be humanely
killed, because keeping it alive will yield no further valid information.
Similarly, if the scientific objective has been achieved, there is no useful
purpose in keeping the animal in further pain and distress. If, however,
an animal is suffering but is still expected to yield useful information, it
may be possible to alleviate the animal's condition and so reduce the level
of suffering through the use of analgesics or sedatives. On the other hand,
it must always remembered that the level of suffering may have become
incompatible with the expected degree of scientific benefit, and so, on an
ethical harm-benefit analysis, the animal should be killed.

In the past it was not uncommon to use death as an end point. However, death is rarely related to the experimental variable under study but rather to indirect effects such as dehydration and starvation by animals' not being able to drink or eat. Dehydration leads to hemoconcentration and an increased viscosity of the blood that the heart cannot cope, which leads to heart failure. Inadequate food intake in rodents can lead to low body temperatures and death. Because these steps to death can take several days, surrogate lethal end points need to be established. One approach is to closely observe the clinical signs preceding death and determine those signs that are shown to be irrevocably linked with death and then use such signs as prelethal surrogate end points. The idea of using early clinical signs to predict later ones requires validation studies in which it is shown that animals will normally progress in that way and that such an end point is reliable. This approach can be used in a variety of experiments, including toxicity testing, the evaluation of medicines such as in vaccine potency testing, the protection test for novel antibiotics, virulence assessments for microorganisms or parasites, batch testing of natural or synthetic products, and so on. Cussler, Morton, and Hendriksen (1999) present some data on a lethal rabies vaccine potency test in which it was found that vaccinated mice given challenge doses of virus went through a series of predictable clinical signs. The authors report that animals showing slow circular movements invariably progressed to death—the traditional end point for the test—and could be reliably used as the end point instead. Soothill, Morton, and Ahmad (1992), in an investigation into the effectiveness of phages for resistant staphylococci, showed that animal suffering could be reduced by several hours by taking a body temperature of less that 35 degrees Celsius as a prelethal end point.

Severity Limits and Their Implementation

In the United Kingdom, under the Animals (Scientific Procedures) Act of 1986, each scientific procedure (e.g., an animal model for a human disease) has a severity limit that should not be exceeded, and indeed it is a criminal offense to do so. The issue is how that limit is recognized in practical terms. While in theory there are four recognized severity bands— mild, moderate, substantial, and severe—severe pain or severe distress is not permitted under any circumstances (interestingly, no scientific case has been sufficiently compelling to justify it). In a sense, what the bands are called is irrelevant; what matters is how they are interpreted. To interpret them accurately and reproducibly requires careful observation of animals, and score sheets documenting clinical signs with time are invaluable (see the following section and Morton 1985, 1986, 1990, 1994, 1995b, 1997a,c, 1998a,b, 2000; Morton and Griffiths 1985; Morton and Townsend 1995). Severity limits can be interpreted as the degree of deviation from normality coupled with other indicators of health and quality of life. Taking body weight as an example, a body weight loss of up to 10, 20, or 25 percent, or greater than 25 percent compared with controls can be used to interpret mild, moderate, substantial, and severe limits, respectively. However, body weight alone may not be adequate, because animals with tumors or ascites may increase in body weight but lose body condition (i.e., muscle mass) and be experiencing extreme suffering. Alternatively, animals that have a body-weight loss of 25 percent and that are diabetic or that have exocrine pancreatic deficiency may be very lively and have a good quality of life. It is important, therefore, that a holistic approach be taken as outlined in the next section so that clear clinical signs can be used to determine humane end points in accordance with the scientific benefit and humane research.

Score Sheets: An Approach to the Recognition and Assessment of Animal Suffering

Many laboratories are approaching the recognition and assessment of suffering by using score sheets that list cardinal clinical signs observable and measurable in animals undergoing a particular scientific procedure. Score sheets are developed through experience and are generally unique to the husbandry and to the specific experiment, as well as to the species and even the breed or strain of animal being used. It is not possible to make a general score sheet for all experiments or for all species. One only has to consider the different potential adverse effects of a skin graft compared with a kidney or heart transplant to appreciate the different signs that might occur (e.g., hemorrhage, rejection of skin compared with the rejection of a dependent kidney or an accessory heterotopic heart in the abdomen). Obtaining professional advice by seeking the opinions of experienced laboratory-animal veterinarians, animal technicians, and stock persons can be very helpful in determining what clinical signs should be anticipated. One initially develops a list of signs by observing the first few animals undergoing a novel scientific procedure very closely and then modifying the list with experience until one obtains a set of cardinal signs that most animals will show during that experiment and that are simple to observe and relevant to the assessment of suffering. These key clinical signs are set out against time in the score sheet (see table 2). On the left-hand column are listed clinical and behavioral signs and along the top are listed the days and the time of the recorded observations. The method of scoring is to record clinical signs as only being present or absent, as indicated by a plus or a minus sign (or sometimes both signs [+ / −] if the observer is unsure). According to convention, negative signs indicate normality or within the normal range and positive signs indicate compromised animal well-being. In this way it is possible to visually scan a score sheet to gain an impression of animal well-being: the more plusses, the more an animal has deviated from normality, with the inference that it is suffering more than earlier. Clinical treatments and other observations are also recorded. Animals can be scored at any time, and it would be certainly more than once daily during critical periods when animals could predictably give rise to concern (e.g., in the immediate postoperative period or in a study on infection after the incubation period).

Practically, it is important to develop a disciplined approach and strategy to recognize adverse effects in animals. At the beginning of an assessment, the animal should be viewed from a distance, with its natural undisturbed behavior and its appearance noted. Next, as the observer approaches the pen or removes the cage lid, the animal will inevitably start to interact with the observer, and its response can be used to determine whether it is normal. Finally, a detailed clinical examination can be carried out by handling and restraining the animal in some way and observing its appearance carefully, in addition to making any relevant clinical measurements (e.g., body weight, temperature) and noting its behavior (it may have become more aggressive or fearful, or even vocalize).

At the bottom of the sheet are guidance notes for animal caretakers concerning what they should provide in terms of husbandry and care for animals undergoing the particular scientific procedure. Also included are guidelines on how to score qualitative clinical signs (such as diarrhea and respiration), as well as the criteria by which to judge humane end points. Finally, if an animal has to be killed, instruction is included about what other actions should be taken (such as tissues to be retrieved or placed in

TABLE **2**

Animal Score Sheet (Blank) for Streptozotocin Diabetes Model

Rat No.		Animal Issue No:				
Date of Starving:		Prestarved Weight:				
Date						
Day						
Time						
From a Distance						
Inactive						
Isolated						
Walking on tiptoe						
Hunched posture						
Starey coat						
Type of breathing*						
Grooming						
On Handling						
Not inquisitive and alert						
Not eating						
Not drinking						
Body weight (g)						
Percent change from start						
Body temperature (degrees C)						
Pale or sunken eyes						
Dehydration						
Diarrhea 0 to 3 (+m or +b)**						
Distended abdomen/swollen						
Vocalization on gentle palpation						
Nothing abnormal detected (NAD)						

Given 5ml saline s/c or p.o.						
Other signs noted						
Signature:						

Special husbandry requirements
Animals should be kept in grid cage with tray and cleaned twice daily, and mouse box for enrichment. Two bottles should be provided for each cage and filled twice daily.
Deprivation of water overnight may be sufficient to cause death by dehydration.
Autoclaved diet must be provided.

Scoring details
*Breathing: R = rapid; S = shallow; L = labored; N = normal
**0 = normal; 1 = loose feces on floor; 2 = pools of feces on floor; 3 = running out on handling; + 'm' = feces contain mucus; + 'b' = feces contain blood

Humane end points and actions
1. *Any animals showing signs of coma within the first 24–48 hrs will be killed.*
2. *Any animals weighing less than the starting weight after 7 days will be killed.*
3. *Any animal showing tiptoe or slow ponderous gait will be killed.*
4. *Inform scientist, named veterinary surgeon, and animal technician in day-to-day care if any of 1 to 3 above are seen.*

Scientific measures
Animals that have to be killed should have their kidneys placed in formal saline and the pots clearly labeled.

formal saline). This helps ensure that the maximum information is always obtained from any animal in the study.

While these sheets take time to fill in, it is not difficult for an experienced person (such as a veterinary technician) to see if an animal is unwell; thus the time taken can be reduced by simply scoring that the animal is normal by marking the NAD box (nothing abnormal detected). However, if an animal is not normal, it does take time for the person to check it and to make judgments over what actions are to be taken. But is that not the price for practicing humane science?

In order to promote good care and good continuity of care, laboratories should allocate an animal technician to be responsible for liaising with the scientists and other technical staff, as well as to maintain and update the score sheets. The roles of the technician in charge are as follows:

- to check that the appropriate licenses and approved protocols are in order and to cross-check them with what the scientist actually intends to do that day to the animal(s);

- to verify that the score sheet is appropriate before an experiment begins;
- to know the purpose of the experiment and the scientific objectives, and to become familiar with the scientific procedures to be carried out on the animals and the clinical signs that may occur;
- to ensure all personnel (technicians, scientists) know how to use score sheets and can recognize the clinical signs and interpret the signs clearly into humane end points;
- to check that technicians not familiar with that experiment, for instance, doing a weekend or holiday rota, are informed about animals;
- to liaise with licensees over the experiment (e.g., timing, numbers of animals, equipment, end points);
- to update the score sheets based on new signs or combinations of signs observed; and
- to report to the responsible persons (e.g., the director, veterinarian researcher) any concerns over the animals or personnel involved.

Table 3 shows a completed score sheet from a real case study. At a glance one can see that there are more plusses to the right than to the left. Several other points can also be noted. First, along the top of the sheet, one can see that as the animal became unwell, it was scored more frequently. During day 0 (the day of the operation) it scored abnormal in one or two predictable signs as it was recovering from the anaesthetic and the surgery (low body temperature and hunched), and so the NAD box was marked. The next day (21 June) basic observations were made of amount of the food eaten, temperature, and body weight, and again the NAD box was checked. However, toward the end of that day, the coat became starey (ruffled), the body temperature rose, and the breathing became more rapid. By the next morning there was a significant body-weight loss (12 percent), which increased during the day to 18 percent—a strong indication that the animal had not eaten or drunk much, if anything, and that it probably had diarrhea. In fact, there were so many abnormal clinical signs that it was decided to kill the animal on humane grounds before the end of the experiment. The sudden appearance of diarrhea and the concomitant rapid weight loss and dehydration, labored breathing, abnormal posture, lack of a red-light response, and so on all confirmed that the animal was becoming severely physiologically compromised and would not yield valid results in relation to the scientific objective. Even more significant, its temperature was now at 35.1 degrees Celsius—a very poor sign—and the extremities were blue (i.e., the color of the feet and ears). In our experience, this animal would have died that night if not sooner.

This scheme of scoring clinical signs for the recognition and assessment of adverse effects on animals during scientific procedures has been shown to have several advantages, which include the following:

- Closer observation of animals is now carried out by all staff at critical times in the experiment because the sheets have indicated the times when animals find their circumstances most aversive.
- Disputed subjective assessments of suffering by staff and scientists are avoided, thereby promoting more fruitful dialogue. Evidence-informed opinion becomes available based on the clinical signs. In a sense the sheets empower the technicians by helping them illustrate to less-experienced persons why an animal is "not right."
- Recording clinical signs adds further measurable scientific outcomes to an experiment; in this way extra useful data are made available, some of which may correspond closely to clinical signs observed in humans with that "disease."
- Consistency of scoring is increased because the guidance is clear and the scoring options are limited.
- Single signs or combinations of signs can be used to indicate overall severity of the procedure, as well as alleviative therapies or scientific procedures at specific points in an experiment (e.g., blood sampling).
- The sheets help determine the effectiveness of any therapy intended to relieve adverse effects.
- The sheets can be used to determine which experimental models cause the least pain and distress, for example, by comparing alternative animal models, thus helping to refine scientific procedures.
- The sheets help in training those inexperienced in the assessment of adverse effects or in that particular scientific procedure.

The score-sheet system provides a visual aid, opens up discussion between interested parties, and helps focus attention on the condition of the animals throughout the procedures. An analysis of the score sheets can reveal patterns of recovery or deterioration and so gives a better picture of the effect of a procedure on the animals from start to finish. The sheet encour-

TABLE 3

Animal Score Sheet (Completed) for Heterotopic Kidney Transplant

Rat No. HN1   Animal Issue No: 234

Date of Operation: 20th June at 11.00hr   Pre-operation Weight: 250G

Date	20	20	21	21	22	22	22
Day							
Time	13.30	17.30	8.00	16.00	8.00	11.00	14.00
From a Distance							
Inactive	2			2	2/1	1	1
Inactive? Try red light response					2	1	1
Isolated	2			2	1	1	1
Hunched posture	1			1/2	2/1	1	1
Starey coat	2			1	1	1	1
Rate of breathing	54			60	64	70	40
*Type of breathing					R	R	L
On Handling							
Not inquisitive and alert	2			2	2	2	1
**Eating jelly mash? Amount eaten	2		50%	2	?	?	?
Not drinking	2			2	?	?	?
Body weight (g)	254		260	250	221	215	205
Percent change from start	2		6%	0%	212%	214%	218%
Body temperature (degrees C)	35	36.5	37	38	38	36.5	35.5
Crusty red eyes/nose	2			2	2	1	1

Excessive wetness on lower body	2			2	2	2	1
Sunken eyes	2			2	2	2	2
Dehydration	2			2	2	1	1
Coat/wet soiled	2			2	2	2	1
Pale eyes and ears	2			2	2	2	2
Blue extremities	2			2	2	2	1
Stitches okay?/ Date removed:	2			2	2	2	2
***Swelling of graft	2			2	2	2	2/1
NAD		√	√				
Dosing							
Other Diarrhea						1	1
Signature:							

Special husbandry requirements
Animals should be put on a cage liner with tissue paper and a small piece of VetBed.

Scoring details
*Type of breathing: R = rapid, S = shallow, L = labored, N = normal.
**Eaten/Jelly mash – amount? – Record as 0/25/50/75/100%.
***Swelling score 0 = Normal 4+ = rejection and large swelling.

Humane end points and actions
1. Weight loss of 15% or more, inform the investigator, veterinarian, and technician-in-charge.
2. Premoribund state (indicating a failing graft).
3. Any major clinical sign recurs after 24 hrs (marked ^, less than 35°C).
4. Inform scientist, named veterinary surgeon, and animal technician in day-to-day care if any of 1 to 3 above are seen.

Scientific measures
Take 1ml of blood and urine, if possible; place at 4°C.
Place transplanted kidney into 10ml formal saline.

ages all involved to observe the behavior of animals and to recognize normal and abnormal behaviors, which will help in determining animals' responses to various procedures. This in turn will help us to devise ways of refining experimental technique by highlighting the type and timing of any adverse effects. Laboratories that use this system constantly develop and update the sheets with further experience. It is surprising how the process never seems to stop as new staff pick up new signs, or new signs develop as the experimental model is slightly modified. Staff also start to perceive patterns of adverse effects that, when taken as a whole, indicate early death or early deterioration sufficient to warrant the animal being killed on scientific grounds alone. Such information has led to better animal care and has also provided useful scientific information, such as the recognition of neurological deficits, times of epilepsy or weight loss, as well as unexpected findings such as urinary retention in a model of renal failure. Furthermore, by picking up signs of poor animal well-being early, we can implement humane end points sooner and so avoid animals being inadvertently lost from an experiment through unexpected death. In the United Kingdom, where severity limits are imposed on every scientific procedure, the sheets can be used to indicate when such limits have been breached, or are about to be breached, or may have to be reviewed (by the precise observation of the clinical signs).

The scoring system has proved to be especially useful with new procedures or when users are not sure of what effects a procedure will have. In my experience the literature rarely records adverse effects on the animals, or how to avoid or measure them, and I believe scientists have a moral obligation to do so (Morton 1992). Our facility now looks more closely at ways of improving our perioperative care, and in some experiments we have found that recovery is slower than it could be if we used different anesthetics or analgesics, or different intraoperative procedures such as maintaining body temperature or giving a bolus of warm saline before recovery. We try to operate early in the day so animals have maximum time under close observation and can be given more support, such as fluid therapy or special diets (e.g., jelly, fruit, vegetables). This has proven to save animals' lives as well as improve the speed of recovery.

There is a philosophical debate about whether it is better to cause more suffering to fewer animals or less suffering for many to achieve the scientific end. That situation is not common in practice, but the U.K. law takes the view that the level of individual suffering is what matters and thus harms should always be minimized. If an animal is used for a scientific purpose more than once (e.g., raising antibodies, removing blood, multiple surgery), then it may or may not adapt to the procedures. For example, animals do not habituate to a daily intraperitoneal injection of saline after five days (Wadham 1996), and the experience may cause greater fear and anxiety in those animals than if using naive animals. Higher levels of suffering have to be justified on scientific grounds and not on the saving of other animal lives. In one sense one can see greater overall harm caused by inflicting severe pain than by killing animals (i.e., quality as opposed to quantity of life). It is interesting that a similar debate is going on in considering euthanasia for humans.

Improved and New Technologies

The advances in biochemistry through the development of more sensitive assays and the use of smaller volumes of material have had an impact on the volumes of body fluids that need to be collected from animals, as well as on the number of animals needed in a group. Moreover, the development of new technologies has had a similar effect (e.g., CT scanners, whole-body gamma counters, nuclear magnetic resonance, ultrasound, electron spin resonance, and microsurgery) through minimizing the invasiveness of procedures.

Concluding Remarks

Animals can suffer during husbandry, social groupings, breeding conditions, handling, restraint, transport, euthanasia, and so forth, and it is important that these are minimized as far as possible for reasons of good welfare as well as good science. But research is needed into the implementation of the Threes Rs, especially refinement. Grant-awarding bodies should be aware of their responsibilities not only to review carefully the work they fund but also to fund research into refinement. Scientists must also write up their work fully and promptly in order to avoid others' repeating the work. Some may point out that the intensive monitoring of animals I have described here adds to the cost of the research, and of course it does. But we have a moral duty to inflict as little suffering as possible, and such refinement is the price we should be willing to pay as a compassionate society and as humane scientists.

References

Blom, H.J.M., G. Van Tintelen, V. Baumans, J. Van Den Broek, and A.C. Beynen. 1995. Development and application of a preference test system to evaluate housing conditions for laboratory rats. *Applied Animal Behavior Science* 43:279–90.

Brain, P.F. 1992. Understanding the behaviors of feral species may facilitate design of optimal living conditions for common laboratory rodents. *Animal Technology* 43:99.

Bruce, R.D. 1985. An up-and-down procedure for acute toxicity testing. *Fundamental and Applied Toxicology* 6:151–57.

Chamove, A.S. 1996. Reducing animal numbers: Sequential sampling. *Animal Welfare Information Center* 7:3–6.

Claassen, V. 1994. *Neglected factors in pharmacology and neuroscience research: Biopharmaceutics, animal characteristics, maintenance, testing conditions.* Vol. 2 of *Techniques in the behavioural and neural sciences.* Amsterdam: Elsevier Science.

Cussler, K., D.B. Morton, and C.F.M. Hendriksen. 1999. Humane endpoints in vaccine research and quality control. In *Humane endpoints in animal experiments for biomedical research: Proceedings of the international conference 22–25 Nov. 1998,* edited by C.F.M. Hendriksen and D.B. Morton, 95–101. London: Royal Society of Medicine.

Erb, H.N. 1990. A statistical approach for calculating the minimum number of animals needed in research. *Institute of Laboratory Animal Resources News* 32:11–16.

———. 1996. A non-statistical approach for calculating the optimum number of animals needed in research. *Lab Animal* 25:45–49.

Festing, M.F.W. 1994. Reduction of animal use: Experimental design and quality of experiments. *Laboratory Animals* 28:212–21.

Hendriksen, C.F.M., and J. van der Gunn. 1995. Animal models and alternatives in the quality control of vaccines: Are *in vitro* methods or *in vivo* methods the scientific equivalent of the emperor's new clothes? *ATLA* 23:61–73.

Khamis, H.J. 1997. Statistics and the issue of animal numbers in research. *Contemporary Topics* 36:54–59.

Mann, M.D., D.A. Crouse, and E.D. Prentice. 1991. Appropriate animal numbers in biomedical research in light of animal welfare considerations. *Laboratory Animal Science* 41:6–14.

Mead, R. 1988. *The design of experiments.* Cambridge: Cambridge University Press.

Medical Research Council. 1998. *MRC guidelines for good clinical trials.* (Booklet.) Mitcham, Surrey, England: Medical Research Council.

Mellor, D. J., and D. B. Morton. 1997. Humane endpoints in research and testing: Synopsis of the workshop. In *Animal alternatives, welfare, and ethics,* edited by L. F. M. van Zutphen and M. Balls, 297–99. Amsterdam: Elsevier Science.

Morton, D. B. 1985. Some areas of concern in the care and use of experimental animals. Paper presented at the first British Veterinary Association Animal Welfare Foundation Symposium.

———. 1986. The recognition of pain, distress, and discomfort in small laboratory mammals and its assessment. Paper presented at the second British Veterinary Association Animal Welfare Foundation Symposium.

———. 1990. Adverse effects in animals and their relevance to refining scientific procedures. *ATLA* 18:29–39.

———. 1992. A fair press for animals. *New Scientist* 1816 (April 11): 28–30.

———. 1994. The need to refine animal care and use. In *Alternative methods in toxicology and the life sciences: Proceedings of the world congress on alternatives to animal use and the life sciences: Education, research, and testing,* edited by A. M. Goldberg and L. F. M. Zutphen, 35–41. New York: Mary Ann Liebert.

———. 1995a. The post-operative care of small experimental animals and the assessment of pain by score sheets. In *Proceedings of animals in science conference perspectives on their use, care, and welfare,* edited by N. E. Johnston, 82–87. Melbourne: Monash University.

———. 1995b. Recognition and assessment of adverse effects in animals. In *Proceedings of animals in science conference perspectives on their use, care, and welfare,* edited by N. E. Johnston, 131–48. Melbourne: Monash University.

———. 1997a. Ethical and refinement aspects of animal experimentation. In *Veterinary vaccinology,* edited by P. P. Pastoret, J. Blancou, P. Vannier, and C. Verscheuren, 763–85, 853. Amsterdam: Elsevier Science.

———. 1997b. Physiological and philosophical aspects of animal consciousness and self-awareness. In *Building healthy babies: The importance of the pre-natal period: Supplementary papers on "Psychological and neurological development in the human and animal pre-born."* N.P.: Women and Children's Welfare Fund.

———. 1997c A scheme for the recognition and assessment of adverse effects. In *Animal alternatives, welfare, and ethics,* edited by L. F. M. van Zutphen and M. Balls, 235–41. Amsterdam: Elsevier Science.

———. 1998a. The recognition of adverse effects on animals during experiments and its use in the implementation of refinement. *Ethical approaches to animal-based science: Proceedings of the joint ANZCCART/NAEAC conference, Auckland, New Zealand, 19–20 September 1997,* 61–67. Wellington, New Zealand: Australian and New Zealand Council for the Care of Animals in Research and Teaching.

———. 1998b. The use of score sheets in the implementation of humane end points. *Ethical approaches to animal-based science: Proceedings of the joint ANZCCART/NAEAC conference, Auckland, New Zealand, 19–20 September 1997,* 75–82. Wellington, New Zealand: Australian and New Zealand Council for the Care of Animals in Research and Teaching.

Morton, D.B. 2000. A systematic approach for establishing humane endpoints. *ILAR Journal* 41(2): 80–86.

Morton, D. B., and P. H. M. Griffiths. 1985. Guidelines on the recognition of pain, distress, and discomfort in experimental animals and a hypothesis for assessment. *Veterinary Record* 116:431–36.

Morton, D. B., and P. Townsend. 1994. Practical assessment of adverse effects and its use in determining humane end-points. In *Welfare and Science: Proceedings of the Fifth Symposium of the Federation of European Laboratory Animal Science Associations,* edited by John Bunyan, 19–23. London: Royal Society of Medicine Press.

Morton, D. B., and P. Townsend. 1995. Dealing with adverse effects and suffering during animal research. In *Laboratory animals: An introduction for experimenters,* rev. ed., edited by A. A. Tuffery, 215–31. Chichester, England: Wiley.

Russell, W. M. S., and R. L. Burch. 1959. *The principles of humane experimental technique.* London: Methuen. Reprint 1992, Universities Federation for Animal Welfare.

Smith, J. A., and K. Boyd, eds. 1991. *Lives in the balance: The ethics of using animals in biomedical research: The report of the working party of the Institute of Medical Ethics.* Oxford: Oxford University Press.

Soothill, J. S., D. B. Morton, and A. Ahmad. 1992. The HID50 (hypothermia-inducing dose 50): An alternative to the LD50 for measurement of bacterial virulence. *International Journal of Experimental Pathology* 73:95–98.

van den Heuvel, M. J., D. G. Clark, R. J. Fielder, P. P. Koundakjian, G. J. A. Oliver, D. Pelling, N. J. Tomlinson, and A. P. Walker. 1990. The international validation of a fixed dose procedure as an alternative to the classical LD50 test. *Food Chemical Toxicology* 28:469–82.

van Zutphen, L. F. M., V. Baumans, and A. C. Beynen. 1993. *Principles of laboratory animal science: A contribution to the humane use and care of animals and to the quality of experimental results.* Amsterdam: Elsevier Science.

Wadham, J. J. B. 1996. Recognition and reduction of adverse effects in research on rodents. Ph.D. dissertation, University of Birmingham.

Workman, P., P. Twentyman, F. Balkwill, A. Balmain, D. Chaplin, J. Double, J. Embleton, D. Newell, R. Raymond, J. Stables, T. Stephens, and J. Wallace [United Kingdom Co-ordinating Committee on Cancer Research]. 1998. Guidelines for the welfare of animals in experimental neoplasia (revised July 1997, 2nd ed.). *British Journal of Cancer* 77:1–10.

The Editors

John P. Gluck is a professor of psychology at the University of New Mexico and a senior research fellow of the Kennedy Institute of Ethics, Georgetown University. He is also director of the Research Ethics Service Project at the University of New Mexico, which provides curriculum and ethics consultation to researchers using both human and animal experimental subjects. His early work involved the assessment of the effects of early experience on learning and animal models of psychopathology. More recently he has written on the ethical training of scientists and the implications of the analysis of moral standing of animals.

Tony DiPasquale is a senior-level graduate student at the Department of Clinical Psychology at the University of New Mexico. He has been involved in various aspects of animal and human research activities since 1984. In addition to the study of research ethics, other interests include ethical concerns inherent in end-of-life issues, evolutionary psychology, and clinical assessment and the delivery of high-quality psychological services.

F. Barbara Orlans is a senior research fellow at the Kennedy Institute of Ethics, Georgetown University. She is the founder of the Scientists' Center for Animal Welfare and a nationally renowned expert on the ethics and policy issues of animal experimentation. She is author of the award-winning book *In the Name of Science: Issues in Responsible Animal Experimentation* (1993) and senior author of *The Human Use of Animals: Case Studies in Ethical Choice* (1998). In previous years she conducted animal experiments on heart disease at the National Institutes of Health.

Contributors

Marc Bekoff, Ph.D., is professor of biology at the University of Colorado, Boulder, and is a Fellow of the Animal Behavior Society and a former Guggenheim Fellow. He recently has been awarded the Animal Behavior Society's Exemplar Award for major long-term contributions to the field of animal behavior. His main areas of research include cognitive ethology, behavioral ecology, and the welfare of animals in research. He has published over one hundred fifty papers and thirteen books.

Nikola Biller-Andorno, M.D., Ph.D., is junior professor of medical ethics at the University of Göttingen, Germany, and a member of the German Academy of Ethics and Medicine. She has held Visiting Fellow appointments in the United States at Yale University, Harvard University, and most recently at the National Institutes of Health. Her current research interests include the role of empathy and care in bioethics, ethical issues in transplant medicine, and international perspectives in bioethics.

Raymond G. Frey, Ph.D., is professor of philosophy at Bowling Green University and Senior Research Fellow at the Kennedy Institute of Ethics, Georgetown University. He is an internationally respected applied ethicist, with numerous writings on animals and animal welfare issues that have been published both in philosophy and scientific journals. He is the author of the important book *The Case against Animals,* published by Clarendon Press.

Anita Guerrini, Ph.D., is associate professor of history and environmental studies at the University of California, Santa Barbara, and chair of its Institutional Animal Care and Use Committee. Her research on the history of science and medicine has been supported by the National Science Foundation, the French Centre National de la Recherche Scientifique, and the Center for Humanities, Oregon State University. She is the author of many articles and three books on the history of science and medicine, including the forthcoming *Animal and Human Experimentation: A History,* published by Johns Hopkins University Press.

David Morton, Ph.D., MRCVS, is head of the Centre for Biomedical Ethics, Division of Primary Care, Public and Occupational Health, and the director of the Biomedical Services Unit at the University of Birmingham, England. Professor Morton, a senior veterinary surgeon, has been a central international figure in the conduct of research on the recognition of pain and distress in nonhuman animals and its application to experimental methodology. His most recent work on the evaluation of standard methods of humane euthanasia is both exceptionally important and controversial.

F. Barbara Orlans, Ph.D., earned her doctorate in physiology from London University, England. She initially came to the United States to conduct animal research at the National Institutes of Health. In 1979 she founded the Scientist Center for Animal Welfare, where she served as president and director. In 1989 she joined the Kennedy Institute of Ethics, Georgetown University, where she is currently a Senior Research Fellow. Her recent books include *In the Name of Science: Issues in Responsible Animal Experimentation* (sole author) and *The Human Use of Animals: Case Studies in Ethical Choice* (co-author), both of which were published by Oxford University Press.

Bernard E. Rollin, Ph.D., is University Distinguished Professor of Philosophy at Colorado State University, where he is also professor of biomedical sciences, professor of animal sciences, and University Bioethicist. Rollin is the author of twelve books and over two hundred articles that cover a wide range of issues in bioethics. He is a principle architect of U.S. federal laws chartering animal care and use committees and is considered the father of veterinary medical ethics.

Index